A practical guide on quality management in spinning

A practical guide on quality management in spinning

B. Purushothama

WOODHEAD PUBLISHING INDIA PVT LTD
New Delhi • Cambridge • Philadelphia

Published by Woodhead Publishing India Pvt. Ltd.
Woodhead Publishing India Pvt. Ltd., G-2, Vardaan House, 7/28, Ansari Road
Daryaganj, New Delhi – 110002, India
www.woodheadpublishingindia.com

Woodhead Publishing Limited, 80 High Street, Sawston, Cambridge,
CB22 3HJ UK

Woodhead Publishing USA 1518 Walnut Street, Suite1100, Philadelphia
www.woodheadpublishing.com

First published 2011, Reprinted 2017, 2018
© Woodhead Publishing India Pvt. Ltd., 2011
Reprinted 2020

Woodhead Publishing India Pvt. Ltd. ISBN: 978-93-80308-08-1
Woodhead Publishing India Pvt. Ltd. EAN: 9789380308081

Woodhead Publishing Ltd. ISBN: 978-0-85709-006-5

Typeset by Sunshine Graphics, New Delhi
Printed and bound by Replika Press Pvt. Ltd.

Contents

9   Role of technicians in quality management   205

Preface

Quality is demanded by all and customer of spinning is no exemption. The quality has dimensions of product parameters, timely delivery and affordable price and also after sales service. To produce quality is not one man's job. It is a combined effort of all in the organization and also of the customers.

In my experience of 38 years in various spinning mills, in various positions from Production Supervisor to Production Manager, Chief of R&D and Quality Assurance and also in my interactions with various research associations, professional associations etc, I could learn various aspects responsible for getting the quality. In this book an effort is made to recollect them and put in simple possible way so that people in the industry can take advantage. Some of the new concepts of Quality are explained in this book, which are not normally explained in other books.

The book starts with the concepts of quality management system and then the objectives of product. We normally get various norms for quality of yarns, updated from time to time, by various research associations and also from customers. However, they are not end use specific, but are broadly classified as carded and combed, hosiery and weaving yarns etc. However, in reality the customers are interested in the yarns to meet their specific requirements. Therefore, in this book, the concepts of product objectives and the impact of product features at customer's end is discussed with some examples.

To produce good quality, one needs to take action at the source of generation of poor quality. Hence the reasons for getting poor quality is discussed in details with four angles viz. the raw material, the work practices, the machinery conditions and adapting of appropriate technology. The Processwise nonconformities normally observed and the normal complaints from the customers are then discussed. In order to achieve the required results, monitoring the processes with suitable control points and check points are essential. They are discussed in detail.

This book has targeted the shop floor technicians and students willing to become a spinner. The developments in the spinning field in terms of latest machinery are not discussed here, but the concentration is on tuning and maintaining the existing machinery and trying to get best out of it by good work practices. If we do not know how to get best out of the existing machinery and technology, we cannot get success even with latest machineries.

I hope the technicians and the industry get some benefit out of this book and would be able to cater to their customers with the quality they require, so that the industry can progress. I am very much thankful to Woodhead Publishing India Pvt. Ltd., for coming forward to publish this book and encouraging me to write a useful book.

B. Purushothama

1

Introduction to quality management

Quality cannot be inspected into a product; it is either there or not.
It must be bred into the making of the product by the operative; this
is where quality starts. — *Phillip Crosby*

1.1   Introduction

Quality management consists of four parts, viz., quality planning, quality
control, quality assurance and quality improvement. Quality planning deals
with planning the activities to meet the customer needs, whereas quality
control deals with monitoring the activities using different control points
and checks to ensure bad quality does not go to the customer. Quality
assurance is focused on establishing systems and procedures to ensure
that quality is achieved all the time and quality improvement concentrates
on changing needs of the customers and proactively works for improving
the levels of quality not only of the product but also of the systems targeting
reduction in costs, timely services and delivery in time while adhering to
product quality, legal and regulatory requirements and ethical values
(Fig. 1.1).

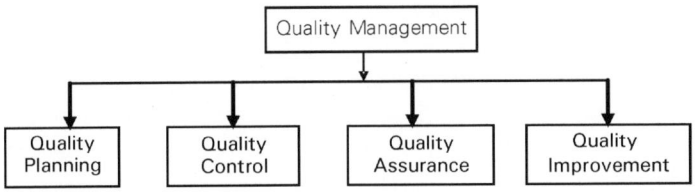

1.1 Components of quality management.

Quality planning is the process of planning the production activities in
order to achieve the goals of meeting the customer requirements in time,
within the available resources. Understanding the customer's needs is the
first step in the process. It is defined in IS/ISO 9000:2000 as "Part of
Quality Management focused on setting Quality Objectives and specifying
necessary operational processes and related resources to fulfill the Quality

Objectives". The product and service provision planning process in ISO 9001:2000 defines the following controls, as appropriate to the product, which is called a quality plan:

- the quality objectives and regulations
- the necessary processes, documents and resources
- the required checks and criteria for product acceptance
- the records needed

In textiles, the ultimate consumers and the men involved in retailing are, normally, not technicians. Therefore, the customer's requirements are not clearly captured and explained as required to a shop floor technician. Although in some cases, technicians are employed for identifying the specific needs, the interpretation changes, and the production personnel get a different message. Over specification is a common phenomenon adapted to ensure reliability, which is resulting in increased expenses. Let us take an example of 20s carded hosiery for knitting purpose. The customers always ask for the best yarn, and often refer to a benchmark like Uster Statistics and demand for 5% or 25% level. They never try to realize whether that quality is required for the product being manufactured and the technology adapted. One should understand that Uster 25% indicates the quality level achieved by the top 25% of the mills who participated in the survey in each parameter separately. Normally the mills with new equipments participate in such surveys and others hesitate. All the mills on earth are not participating. Further, unless Ms. Zellweger Uster requests a mill to participate, the mills will not participate. That being the case, we should try to know what is happening to the yarns made by other mills, which are inferior to Uster 25% level. If that yarn can be used to get the required end product, why should not we use it? Whether the quality of yarn required for different end uses are same, e.g., hand knitting, slow-speed knitting, high-speed knitting, etc., or for the different products with single Jersey, pique, interlock, rib, etc., for T-shirts, socks, hand gloves, sanitary napkin covers, undergarments, Rexene and other coated materials? Another important point to be noted is that Zellweger Uster gives statistics for each parameter separately, but the customers interpret that a particular yarn made by some mill had all the parameters at Uster 5% or 25% level. A mill with old machines, which cannot run at high speed gets a higher elongation compared to modern mills with high-speed working and hence comes in better level like Uster 5% level. So we should think before coming to a conclusion.

In some cases, the customer gives a sample of yarn and asks the spinner to match the quality. The men in laboratory analyze the properties of the yarn sample and give report to the spinner. The spinner tries to keep the same parameters of twist and count at spinning stage. Normally people do not realize that the samples given are not ring frame cops but are cones, hanks or a piece of cloth from which the yarn is taken out.

The yarn properties on cone/hank/fabric are normally different compared to what it was at ring spinning stage. Studies have shown that the count becomes slightly fine as we process the yarn in winding and is normally attributed to tension and some fibre loss during winding. Normally count becomes slightly fine in winding, especially in soft-twisted yarns like hosiery yarns. The count becomes further fine when the yarn is bleached or dyed in light shades, but becomes coarser when dyed with extra dark shades.

The change in twist per metre was also observed as the yarn undergoes winding operation. The twist increases from ring frame cop to winding. This is attributed to the side unwinding. The increase in twist is equal to the number of coils unwound. One can see it by winding a flat tape on a spindle and unwinding from the nose (see Fig. 1.2). In case of side unwinding of bobbins this change in twist is not observed.

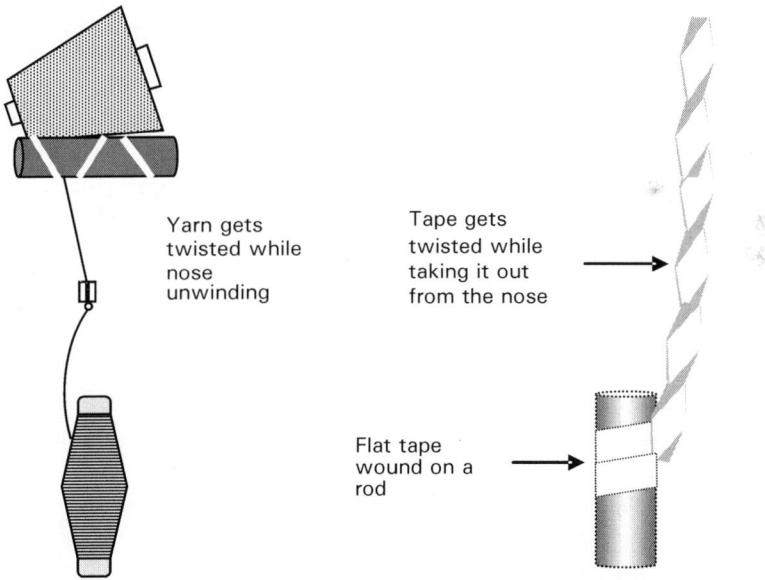

Yarn gets twisted while nose unwinding

Tape gets twisted while taking it out from the nose

Flat tape wound on a rod

1.2 Twisting of yarn.

The increase in TPM matches with the coils per metre of yarn. This is irrespective of the direction of twist. Same is not the case if the cones are rewound. The twist increases while rewinding in case of "Z" twist and reduces in case of "S" twist. If the yarn has undergone the process of warping or knitting, again there shall be an increase in TPM in case of "Z" twist yarns, but a reduction in "S" twist yarns. In case of the yarn undergoing wet processing, it is difficult to assess the correct twist per metre. Hence while deciding the parameters in ring frames this fact is to be kept in mind.

1.2   Understanding customer needs

If we need to be successful in the international market, we need to comply both with the Product Quality Norms and the National and International Trade Regulations. Unless we comply both, it shall not be possible to be in the market. The normal requirements may be listed as follows:

Specified needs

- The average count and the tolerance
- Maximum variation (CV%) tolerated in count
- The twist per metre and the tolerance for TPM
- Maximum allowed variations in twist (CV% of TPM)
- The minimum breaking strength (RKM or CSP) and the maximum allowed variation
- The minimum elongation at break and the maximum allowed variations in elongation
- Maximum permitted unevenness (U%)
- Maximum limits for imperfections per KM – thick, thin and neps.
- The maximum hairiness and variations permitted
- Class-wise maximum permitted faults per 100 KM (Classimat or Classifault)
- Maximum permissible level of contaminations
- The average cone weight and the tolerance

The unspecified needs include uniformity in shade through out the lot and between lots, to absorb size uniformly during sizing, fault-free winding, lower breakages at warping, sizing, looms and in knitting, able to get the required shade of dyeing right at first time in all the lots of dyeing, able to get the required feel of the fabric and so on.

The specific needs of a customer are easy to understand, while we need to work hard to understand unspecified needs, which are much higher compared to specified needs.

The product specific technical parameters include the fibre, the count, carded or combed, for weaving or hosiery, the minimum strength required, the twist range, the permissible U% and imperfections, the cone weight, cone diameter and number of cones per package or per order. We also need to consider the instructions to users by means of care labels, packing and labeling, the other regulations like the markings as per statutory requirements, adhering to the norms of customs, imports and exports, ethical work norms, non-usage of harmful materials and controlling the effluents, etc.

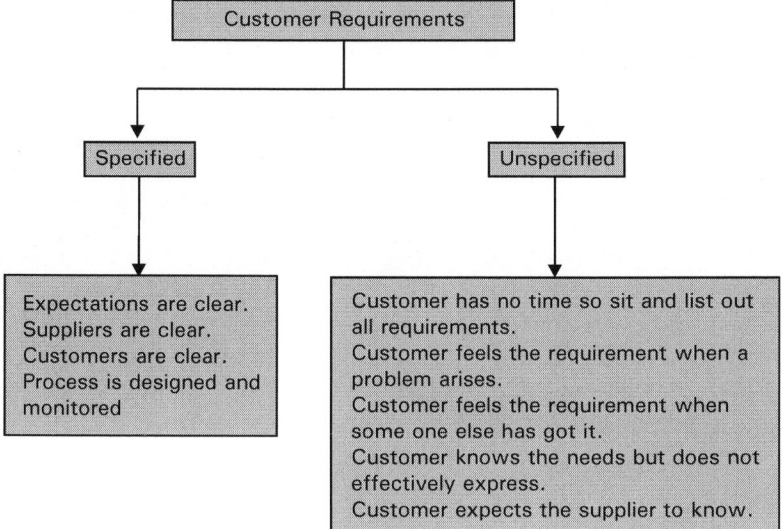

1.3 Customer requirements.

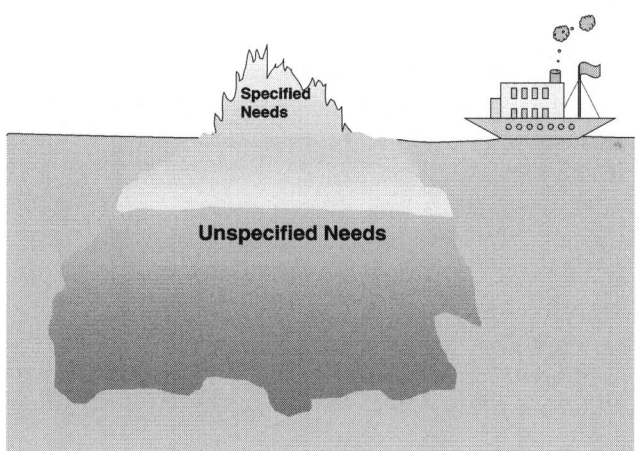

1.4 Specified and unspecified needs of a customer.

Once the parameters needed is clear, one can plan for the raw materials, the processes needed, the machines to work, the process controls needed, the inspection and testing, the training needed for operators, the starting and ending date of production, the infrastructures needed, the information required, etc.

1.3   Quality objectives of the product

Any product manufactured should have an objective. It should be clear to the spinner as to what is expected out of the yarn and the way in which the yarn is supposed to work, how it should contribute for the success of the end product to do its functions or meet the customer needs and expectations. If the quality objectives of the product are clear then it helps in deciding the specifications. Let us take an example of a combed hosiery yarn. When we say combed hosiery people jump to a conclusion of TM say 3.6. But if we discuss as to what purpose the yarn is used; whether for T-shirt, banian, sports socks, baby socks, baby wear, candle wick, sewing thread, industrial application like Rexene, towels, embroidery threads, etc., we get a clear idea as to what TM is needed. For sports socks we need slightly higher TM compared to baby socks. Candle wicks require a lower TM compared to T-shirts. Further, we need to understand the type of machine on which our yarn works. If knitting is done on a slow-speed machine, it can work with a very low TM. If the knitting speed is high, then a slightly higher TM is preferred to have good working and also lower linting. In Chapter 2 the concepts of quality objectives of products are discussed in detail.

To ensure that parameters are met all the time, internal targets are to be fixed at each process by carefully assessing the process capability of the plant in operation; in other words, six sigma concepts. Each mill has to work out internal targets for each type of machine they have. The process control norms also depend on the process and machine capability. Once the targets are fixed, the process and procedures can be established, and resources can be planned and provided. The inspection and test plan depends on the customer requirements and not on the process capability or machine capability, as the customer is least interested on the problems we have, but is concerned about his getting correct material in time. If we give something better than the expectation of customer without increasing the price, he shall be happy, but shall not accept any deviation in quality in further supplies. Hence while deciding the internal norms and acceptance criteria, one should work very carefully. The planning for production should be done in such a way that customer gets what he expected and we do not over do any thing. Therefore, the important steps in quality planning are as follows:

- Understanding customer needs and deciding quality objectives of the product.
- Identification of processes needed.
- Identification of machines to work.
- Determination of process controls and norms.
- Determination of inspection, testing and acceptance criteria.
- Planning the starting and ending time.
- Identification of persons to operate and control.
- Identification of training needs.
- Working out infrastructure requirement.

The plans need to be documented for effective follow up while manufacturing and for monitoring the process. An effective quality plan helps in achieving the results as expected in product and services. It is better to prepare checklists for follow up, so as to ensure that nothing is left out from what was planned.

1.4   Balancing the processes – quality planning

Balancing the machineries includes aspects like getting optimum utilization of the machines, ensuring smooth flow of materials between processes, avoiding stock build-up and starvation due to want of back material, ensuring that the production targets are met as per customer's delivery requirements and at the same time ensuring that the speed and hank organization selected give the quality of yarn as committed to customer. The normal wastes generated in each process, both saleable and usable, needs to be added while working out the productions.

Although the process is balanced and the machineries are allotted as per the calculations, we see unbalancing because of working problems encountered in certain qualities, unforeseen breakdowns, rejection of in-process materials due to quality issues, absenteeism in certain section, etc. The shop floor technicians need to workout alternate plans to take care of such situation. They need to verify the suitability of alternate process to get both quality and productivity. Normally we get quality problems in this type of short-term adjustments, as we might not be able to adapt all parameters as decided. We might have to accept a via-media solution.

1.5   Quality control

A system for ensuring the maintenance of proper standards in manufactured goods, especially by periodic random inspection of the product.   – *www.answer.com*

A system for achieving or maintaining the desired level of quality in a manufactured product by inspecting samples and assessing what changes may be needed in the manufacturing process.

– *msn.encarta.freedictionary*

The quality control is the process of checking and monitoring the process and products with an intention of preventing non-conforming materials from going to the customer. Various result areas are identified for each process and studies are conducted to verify whether those results are being achieved. Normally a separate set of people designated as quality controllers conduct various studies and tests and highlight the deviations. It is the responsibility of production and maintenance people to take actions and correct the deviations. It is normal practice to refer the quality standards.

- Quality control and standards are one of the most important aspects of the content of any job.
- By a quality standard we mean the establishment of the threshold at which level of severity a defect becomes unacceptable, i.e., a fault.
- It is the equivalent of tolerances applicable to measurable factors.

The control section normally has two separate sections: one for testing the product quality at different stages of production and also of final product, normally termed as inspection and testing, and the second one for studying the process, normally called as process control studies.

Normal tests done in a spinning laboratory are as follows:

Cotton – length, Micronnaire, strength, trash, neps level, colour, honeydew content, UV absorbency

Lap – trash%, fibre length, neps

Sliver – trash, neps, U%, hank variation

Roving – U%, hank variation. Some mills have developed system for measuring roving strength similar to Lea strength testing of yarns.

Yarn – count, count variation, TPM, Lea strength, single thread strength, U%, imperfections, Classimat faults, appearance, snarling tendency and hairiness

1.6   Quality assurance

A planned and systematic pattern of all actions necessary to provide adequate confidence that the product optimally fulfils customers' expectations, i.e., that it is problem free and well able to perform the task it was designed for.   – www.dictionary.die.net

Quality assurance concentrates on identifying various processes, their interactions and sequence, defining the objectives of each process, identifying the key result areas and measures to measure the results, establishing the procedures for getting the required results, documenting the procedures to enable everyone to follow the same, educating the people to implement the procedures, preparing standard operating instructions to guide the people on work spot, monitoring and measuring the performance, taking suitable actions on deviations and continuously improving the systems.

The involvement of all in the organization starting from the top management to the lowest worker is essential to assure the quality products and services to customers at all the times. In quality control we check a small sample and take a decision about the quality of the product, and allow it to go to customer, whereas the customer consumes everything, it means the customer checks 100% of the products supplied by actually working it throughout its life. The customers can judge the ability of a company to provide them the required quality in a better way than the in-house quality control people. We need to understand this and make all efforts to assure customers that the products supplied by us perform as per their expectations.

Periodic quality audits are done to ensure that everyone is following the laid out procedures, which is very essential for the consistency in quality. It is essential to do the linking exercises[7] during quality audits to verify organization wide implementation.

1.7   Quality improvement

Systematic approach to reduction or elimination of waste, work-back flow, rework and losses in production process.

– www.businessdictionary.com

All quality improvement occurs on a project-by-project basis and in no other way."   *– Juran*

Quality improvement is a never ending process. The customer's needs and expectations are continuously changing depending on the changes in technology, economy, political situation, ambitions and dreams, competition, etc. One needs to go on analyzing the facts and identify potential problems in advance and take necessary precautions to survive in the competitive world. We need to consolidate all steps we take for improving a situation and implement them uniformly organization wide to get consistent improvement. We need to be always on the toe for facing challenges, benchmark the best and try to re-engineer the activities and the products to beat the competition. The use of Five Golden Questions[1] for self assessment and evaluation of the maturity of implementation of quality management system[2] are essential to have continual improvement in quality.

1.8   Five Golden Questions

The Five Golden Questions[1] explained here are very simple questions, but highly effective if we make sincere efforts to evaluate self and work for improvement. The questions are applicable to all the activities of any organization or person. The questions are as follows:

1. Do we have a procedure?
2. How do we ensure it as the best?
3. How did we implement?

4. Did we get the result as anticipated?
5. How do we compare ourselves with our competitors?

These questions need to be asked again and again in quality management as and when we achieve some results. It will avoid complacency and indicate as to where we need concentration. We should always understand that there is no limit for achieving quality, but the competitor's performance is the judging factor, as to whether our quality is acceptable or not.

Product quality objectives

2.1 Introduction

As already indicated in previous chapter any product manufactured should have an objective. It is true for yarns also. The producer should know the purpose for which the yarn is being used. The spinner should be aware as to what is expected out of the yarn, in what way the yarn is supposed to work, how it should contribute for the success of the end product to do its functions or meet the customer needs and expectations. These are called as quality objectives. If the quality objectives of the product are clear it helps in deciding the specifications.

According to ISO 9000[3], quality objectives are not static and need to be updated in view of the business climate and other continual improvement activities. The quality objectives of a business must be defined; they must reflect the quality policy, be coherent, and align with the overall business objectives, including customer expectations. The quality objectives must have a meaningful result. Similarly the product quality objectives are also to be defined and communicated down the line to the people involved in producing the product, so that they are always alert while producing. This helps in producing the right products right at first time.

Sead Jahic[4] quotes Kotler in his article stating that when a company plans to offer a product to its target group/market, it should take into consideration five levels of product, viz., potential product, augmented product, expected product, basic product and core benefit. In order to set the objectives of quality of products it is also essential to be familiar with "Dimensions of Quality of Products" as given by David A. Garvin. These dimensions are classified by Philip Kotler as the attributes of product differentiation. David A. Garvin suggests different dimensions of quality as:

- Performance
- Characteristics/attributes
- Reliability

- Durability
- Service
- Esthetics
- Tangible quality – adjustments towards standards

When we talk of quality objectives of yarn, the dimensions shall be the working performance on the next machine, the hand and feel of the final fabric, the appearance of the fabric, the durability, comfort of wear, the functional aspects in case of technical application, etc. Sead Jahic explains three phases of setting Product quality objectives as follows:

- Phase 1 – Defining the objectives of quality of the product based on customer demand, preferences, previous experience, organization and positioning strategy
- Phase 2 – Defining the objectives of quality of the inputs/raw materials and setting clear objectives of the quality for the materials depending on internal standards or other means
- Phase 3 – Defining the objectives of quality of the marketing elements considering the three elements of marketing mix: price, distribution and promotion

It is found that for some products the goals of product quality are defined according to previously established standards as ISO, EN or National Standards. However, the business entities have an option, in line with their objectives, to set higher and more demanding standards as well as to establish a scale for new ones. This is normally seen in yarn quality specifications. People want to be competitive and hence try to take the best as benchmark and make efforts to out-beat it.

When new products are being developed, the existing international standards might not be of help for the materials to be procured. We might have to design a new material and give precise specifications. These are truer while designing protective materials, textiles for industrial applications, technical textiles, smart textiles, aviations, etc. We need to understand the requirement of end user and design the product to meet the requirement. If yarns are used as raw material, we need to clearly define the yarn parameters needed to make the final product a success.

In some cases where buyers have matured systems of quality management, and have clear and well defined product quality objectives of products in their portfolio, as well as raw materials they use, a firm can adapt the same directly or define according to its own abilities in agreement with its buyer. In some industries, especially in specialty products manufacturing and processing industries, organizations have their own experiences and GMP, i.e., good manufacturing practices. While setting of objectives for individual processes in a process industry we must take into account the need of all stakeholders. This imposes the linking of quality objectives at all levels and also consider attributes of side effects.

Setting the objectives of product quality is an obligation as per the guidelines of ISO 9001. This is needed for both the products being manufactured and the services being provided. The quality objectives also represent another essential motive to allocate resources and time which is in the interest to any organization. The greatest obstacle in setting these objectives of product quality is multidisciplinary nature of this matter.

By going more in detail, one can work out the quality objectives more precisely and fix the parameters, which are measurable. The quality objectives of a product must address the following:

- For what end use the customer is asking for this material?
- What are the critical quality requirements of the end product?
- What is the intended application of the product with ultimate consumer?
- On what machines our product is likely to work, i.e., high-speed machines, super high-speed machines, slow-speed machines, automatic machines, etc.
- What is the work environment at customer's place, i.e., humidity and temperature, dust level, skill of operatives, the management culture, etc.?
- What are the applicable regulatory requirements for the product from the point of view of product safety, user safety, trade regulations, etc.?

Let us have some discussions on the quality objectives of yarns for different purposes. The discussions in this chapter are complementary to the discussions in Chapter 3, where we discuss the impact of various yarn features at the customer's end depending on the end use. It is suggested for the spinner to prepare the quality objectives for each type of yarn spun by discussing with the marketing personnel and the customers wherever possible.

2.2   Quality objectives of yarns

The quality objectives of the yarn being produced should be specified by the concerned market contact persons and the internal customers of the yarn like winder, weaver, knitter, dyer, etc. The spinners normally have the freedom of selecting the raw materials, adjusting the speeds and settings, bypassing certain machines, etc., but have no say on the specifications and the tolerances. Sometimes the customers specify the raw material to be used, and in such cases, we need to use that material only. Spinner should discuss with the customers, internal as well as external, understand their requirements, prepare a sample and get approved. Afterwards, it should be only matching to the samples, unless otherwise the customer suggests a change depending on the performance at customer's end. Let us discuss some examples.

2.2.1   Combed yarns

Combed hosiery yarns: Combed hosiery yarns are used for various purposes like T-shirts, baby socks, socks for sick and elderly people, banians, sports socks, candle wick, etc. Further they might work on different types of machines at different speed ranges. The yarns supplied by us should perform on the machines that are installed at the customer's end. Hence we need to define minimum strength and strength variations acceptable along with the figures of elongation. The actual count to be maintained depends on the culture of the customer and market. In some markets, people prefer the fabric to be denser, whereas someone wants to produce more length of fabric from the given quantity of yarn. The actual count to be maintained should be decided by discussing with customer and an agreement is to be made. When we talk of socks for babies and elderly people, the softness and comfort becomes more important, and hence we need a softer twist; whereas for sports socks, abrasion resistance is very important. The consistency in colour and dyeability is very important when the yarns are used for external wear like T-shirts, whereas it is not that important when used for banians. The banians are normally made by bleached fabrics that are either knitted using bleached yarns or bleached after knitting. In case of socks, the colour variation between lots of cottons do not create any problem as normally very less number of cones are used to knit socks, like one cone, two cones or four cones. Further the length of fabric in socks is also very small compared to the knitting done on circular knitting machines for T-shirts. Let us discuss the quality objectives of yarns for some specific end uses.

Combed yarns for T-shirts: The T-shirts are outer wear materials, might be used as a casual wear, party wear or for playing certain games like golf. People also use T-shirts made by using combed yarns for morning walking. The appearance is more important where combed yarns are demanded. The count should not be finer than specified as slightly higher GSM is preferred by customers. For this reason, normally in T-shirts, slightly coarser count is accepted but not a fine count. Count variations should be as low as possible. Higher count CV% can lead to barre and streaks that are not liked by the customer. Twist should be slightly more than that is used for banians and baby socks, but lesser than that required for socks. A twist multiplier (TM) of 3.6–3.7 is suggested for T-shirts depending on the speed of working and the cotton used. The fabric should give the feel expected by the customers. Some customers insist of masculine feel whereas some opt for feminine feel. Hence, a spinner needs to discuss this point in detail with the customer. Normally high-speed circular knitting machines are used for producing T-shirts with combed yarns. The yarn should be strong enough to run on those machines. A minimum RKM of 16 is normally demanded. The U% should be low and with lower imperfections. A higher U% with repeated short-term variations can give streaky effect on the fabric. Uniform length of yarn on cones is an important factor to avoid wastage of remnant yarns on cones.

Combed yarns for socks: The socks made by using combed yarns may be used for various purposes like baby socks, executive socks, patient's socks, old people's socks, sports socks, etc. In all cases the count should be as per specification. This is because the weight of socks and also the dimensions vary as per the count. The variation in count should be as low as possible to avoid sock to sock variations in size. Twist should be as low as possible for baby socks, say 3.4–3.5 TM, whereas for elderly people and patients, a TM of 3.5–3.6 should be acceptable. If the socks are for regular office wear a TM of 3.6–3.7 shall be required, and if we want the socks for sports purpose, a TM of 3.7–3.8 is suggested. Normally the tension applied is less in socks knitting compared to T-shirt knitting. Therefore, strength is not a critical requirement. Short-tem unevenness gives a diamond effect on the socks, or a barre effect. The long-term variations result in irregular shape of socks. If the socks have art works like prints or other designs, the imperfections are not critical, but the slubs can create breakages and holes.

Combed yarns for towels: It is a normal understanding that the towel yarns need to have good absorbency. This is true for towels with honey comb weave or Huck-a-back weaves where there shall be one set of warp yarn and one weft. When we talk of Terry towels, there shall be two warp beams; one for base cloth and the other for piles. Nowadays terry towels are being manufactured on high-speed shuttle-less looms. Therefore, the yarn should be strong enough. Hence normal warp yarns with 3.9–4.0 TM are used. In case of yarns going for pile, we need higher absorbency and hence low twist is recommended. For weft, a low twist is always good provided it is strong enough to run on loom. Evenness is not a critical criterion as the designs are seen prominently rather than yarn variations or defects by the customers while selecting a towel. The count is important to maintain the weight of towel, or else it is needed to stabilize the weight by playing with the pile heights, which is not good. It can spoil the complete feel or design of the towel and hence is not accepted.

Combed yarns for warp: The combed yarns are normally used for shirting, sarees, dhotis, dress materials, satins, etc. The warp yarns require higher strength and lower hairiness compared to the yarns used in weft. Normally an RKM of 18 is demanded for combed warp yarns. The count should be as per the agreement with least variation. The twist needs to be given as required by the final product. In case of voiles, we need to give high twist. Sometimes, fabrics made out of combed yarns are used as liner cloth for making Rexene. Uniformity of the yarn becomes a main criterion in such case. After the introduction of compact system, the compact yarns are preferred compared to normal ring spun yarns because of lower hairiness, higher strength and savings in the sizing ingredients. Uniform length on cones is an important factor to avoid wastage of costly combed yarn.

Combed yarns for weft: The combed yarns in weft are used mainly where softness is a criterion. The count normally is tolerated if is slightly coarser, but not finer. The twist becomes a very important factor. It should be as low as possible if the fabric is going for raising application. Again the TPI is an important criterion when the yarn is used for towels. The TM of a weft yarn used for shirting needs be much higher compared to the weft used for towel of raising cloth. The twist in weft also depends on the method of weft insertion used on loom. Very low twist might find it difficult to work on air jet looms, as there are chances of the yarn getting opened up, but the same problem will not come with rapier or gripper looms. Importance of evenness depends on the finish and other processes adapted in wet processing. If the fabric is undergoing printing operation, the imperfections in the weft shall not be critical, but there should be no objectionable faults. Consistency in cotton colour with in lots and between lots is very important to avoid the problems of weft bars. The problem of weft bar is more in combed weft rather than in carded weft.

Combed yarns for embroidery: Normally the combed yarns used for embroidery are doubled, singed, mercerized and dyed. Uniformity in twists and absence of imperfections are demanded. A higher imperfection can result in uneven singeing. People are even ready to pay higher rates for using very superior mixing. For example 100^s combed mixing for yarns of Ne 10^s or 20^s. The strength is not an important factor, but the count is important. Variations in count can lead to cork screw effect which is not accepted. In a number of cases, embroidery yarns are either multifold or cabled with a low twist.

Combed yarns for candle wicks: Candle wicks require a very uniform yarn coarse yarn with low twist. The count variations might not give any problem, but the unevenness is not accepted. The yarn should be clean without any kitties or trash, and hence combed yarns are preferred.

2.2.2   Carded yarns

Carded yarns for hosiery: Carded yarns are used for making T-shirts and socks. Normally for underwear carded yarns are not preferred. If a carded hosiery yarn is taken for producing single jersey on a high-speed knitting machine, the quality objectives demanded shall be something as follows:

- Evenness to get a uniform appearance of the fabric.
- Sufficient and uniform strength to work on high-speed knitting machine without excessive breaks.
- Minimum lint generation while knitting at high speeds.
- Free-form dead cottons and immature fibres to get a uniform depth of dyeing.
- Flexible and smooth to move freely in the needles.
- Uniformity with a low twist to have smooth feel and without spirality.
- Uniform density of cones to have uniform tension and even knitting.
- Equal length of yarn on cones to have lower wastages.
- Count as required to get the GSM as specified by the ultimate customer.

Carded yarns for socks: Carded yarns in socks are used mainly for daily wear, school uniforms, sports wear and in security services. The carded yarns are taken because they are cheap. As carded yarns do not give a soft feel like combed yarns, they are not used for baby socks and for the patient's socks. Short-term unevenness is a major complaint regarding carded yarns, which give diamond type effect on the socks. The count variations result in uneven sizes of the socks, and much time is wasted in pairing the socks. Therefore, the count variations should be maintained as low as possible, and also the average count should be as specified.

Yarns for woven apparels: The apparels need a good appearance, and hence presence of long-term or short-term unevenness, imperfections and faults are not accepted. People go for carded yarns to save the cost, but expect an appearance similar to a combed yarn. Therefore, if the spinner is clear that the yarn is going for apparel end use, can take suitable measures like selecting good cards that give good nep free web, set the blow room to extract the trash and kitties to the maximum extent, take more flat strips than normal, etc.

Yarn for canvas belt: The canvas belts are very heavy fabrics designed to withstand a high tension. However the exact load the belt has to withstand depends on the part it is supposed to drive. To withstand higher tensions, the tensile and elongation properties of the yarn should be very good. Therefore, normally cabled or multifold yarns are used. This belt has to drive flat pulleys of machines and hence it should have a grip. The belt should not wear-out because of abrasion and hence good abrasion resistance is required. Hence, long fibres and slightly higher twist in the yarns are preferred. The pulleys become hot while working, and also the belt. The belt should not deform due to high heat and tension. There might be certain working condition like high humidity and temperatures, fumes of various chemicals, dust, etc., and the belt should perform. The belt consumes more power if it is heavy, and hence, it should be as light as possible, but strong enough to drive the machines as needed. The belt should have uniform surface so that there are no jerks while driving. Hence the yarn should be fairly uniform in diameter. If the yarn has knots, it results in breaks while preparing the belt and results in stoppages. Hence long knotless yarns are preferred.

Yarns for flannels for sizing: The flannel cloth should give a good cushion to the yarn being sized, should hold the size on to its surface and transfer the same to yarn being sized. Hence the twist should be lower compared to the yarns used for making flat belts. The flannel does not undergo stress as belt does, but it undergoes compression. The diameter of the yarn becomes more important for a flannel and higher diameters are always preferred. This is the reason that wool is preferred to cotton. The felt should not wear out because of abrasion; hence longer fibres are preferred.

2.2.3   Miscellaneous yarns

Yarns for parachutes: Parachutes are folded in a small bag and carried on the back of sky divers. Hence it should be light. The parachute opens up while jumping from a plane. At that time, the fabric shall get sudden impact of wind at high speed and pressure; it should not burst. Therefore, the bursting strength of parachutes must be very high. These are folded in a small bag when not in use; hence the fabric thickness should be as low as possible. That is why silk and nylons are preferred and not cotton or wool. The parachute remains in a folded state for a very long time compared to it being in open state; hence the fabric should not get deshaped at creases. Therefore, we need fibres with a good drape and it should not lose its properties if kept in folded condition for a very long time. The parachute should not become wet and heavy by absorbing moisture from atmosphere; hence slightly higher twists are required in the yarn and the yarns should not have hairs on their surface. This is the reason the filaments are used.

Yarns for tyre cords: The tyres undergo heavy compression and abrasion while in use. Because of the weight of the vehicle there is always pressure on the tyres. Depending on the road conditions, the speed in which the vehicles move, the brakes application, etc., there shall be sudden impacts on the tyre. The tyres abrade when brakes are applied. Therefore, the tyre cord should have good tensile properties like high bursting strength, high ballistic strength, good abrasion resistance, good elongation and elasticity, etc. The strength and elongation of yarns are therefore very important criteria. Normally cotton tyre cords are used for cycle tyres, whereas nylon tyre cords are used for other vehicles.

Yarns for sewing: The sewing thread should pass freely in the needle eyes and should not break while the stitching operation is on. Therefore, absence of knots becomes a very important criterion. Unevenness with thick and thin places results in breakages while stitching, and hence the yarns should be as uniform as possible. The yarn should be round in shape. This is the reason for using threefold yarns. The bulk of the yarn should be less while the strength and elongation should be good to have better seam strength. The sewing thread should be free from hairs, as the presence of hairs can hamper the movement of thread in the needle. The sewing threads are normally singed.

2.3   Communications of quality objectives

The quality objectives of the yarns being produced should be clear to the people working on the spot in order to avoid quality complaints and getting the required quality right at first time. The workers and the supervisors need to be trained adequately and informed from time to time regarding the changes in the objectives.

It is a normal practice with leading buyers to send their representative to the mills to study the working conditions and to explain the critical requirements while starting a new product or new order. Sometimes the buyer's representative audits the systems periodically to assess the continued suitability of the mill to supply the yarn. It is essential to communicate the audit findings to all in the work area so that suitable precautionary steps are taken.

Some good mills have a practice of sending their workers to their customers as a routine to understand the performance of their yarn and any change if required in the product. The workers those visit the customer's place shall make a presentation of their observations to the management and fellow workers, so that all can jointly discuss the actions to be taken. The control points and check points are decided and the monitoring of the quality is done suitably. This helps in reducing the quality related market complaints and claims. As the workers and staff interact with the customers, minor problems are sorted out and are not registered as complaints or claims as the customers are more interested in performance rather than on claiming or fighting.

Some of the customers and the suppliers have a mutual understanding that the customer must find minimum five areas of improvement in each consignment and give feedback to the spinner, and the spinner must attend to them and report back to customer. This type of transactions is found specifically where specialty yarns are manufactured for specific performance wears.

Impact of yarn features at customer's end

3.1 Looking from customer's perspective

The yarn quality is normally expressed by the features like count, CV% of count, twist, CV% of twist, strength, imperfections, etc. Lower variation is normally considered as good, but it need not be really good as it depends on the end use. For some applications, the customers demand yarns with higher imperfections. Therefore, a spinner should know the impacts of various parameters for different end uses. Let us discuss the impact of yarn parameters at customer's end depending on the end use for which the yarn is used.

3.2 Knitting applications

3.2.1 Grey yarns for knitting T-shirts

- *Count variation*: A deviation in average counts results in higher or lower weight of fabric, which also affects fabric feel and wear comforts. A coarser count leads to lesser fabric area with given yarn, and hence becomes costly, whereas a finer count gives lesser GSM of fabric, which is not liked by the customer. In case where the count is coarser than the gauge of the knitting machine, it leads to problems in working also. Variations in count within and between cones result in streaky or barre effects.
- *Twist variation*: A shift in twist level results in a change in fabric feel. A higher twist gives harsh feel and also results in swaying of the fabric. A lower twist gives soft feel, but results in lint generation while knitting. A lower twisted yarn can get stretched easily in the knitting creel resulting in thin places. Variation in twist within and between cones results in streaky or barre appearance. Twist variations also can lead to uneven shrinkage of the fabric.
- *Imperfections*: Increase in imperfections and U% leads to a cloudy and streaky appearance. However, in number of cases, a very even yarn is also not liked by the customers as it gives the so called "feminine effect".
- *Strength:* Normally strength is not a major criterion required for

knitting; however, with the introduction of super high-speed knitting machines, the strength is also getting a weightage. A lower strength may result in a lower knitting efficiency due to higher breakages.

- **Tensile variation:** Higher variation in tensile results in poor knitting performance. This also is likely to result in uneven fabric dimension due to differential shrinking.
- **Elongation:** A lower elongation makes the yarn brittle, and resists bending of yarn while passing through different parts of the knitting machine, and creates breakages or uneven fabric. It might result in variations in loop height, resulting in bars.
- **Friction value:** A higher friction value resists the smooth flow of yarn while passing through metal surfaces like needles and hooks, results in excessive breaks and poor performance in knitting.
- **Hairiness:** An increase in hairiness gives a fuzzy appearance of garments. However, some customers like this effect, and some other do not. Variations in hairiness between cones or between cops in a cone give barre effect.
- **Winding quality:** Uneven winding results in uneven tension in knitting, giving rise to uneven tightness factor and horizontal lines and barre effect. If the cones have stitches in the back, it results in breakages. The front stitches leads to loose ends, but is not critical as it can be controlled by the tensioning device at knitting.
- **Consistency in cotton shade and Micronnaire:** Day to day variations in cotton shades and Micronnaire values between mixings in the same lot of yarns results in uneven dyeing, streakiness and barre. Sometimes the defect can be seen only after dyeing the fabric that too in particular shades.
- **Objectionable faults:** The objectionable faults, normally slubs, bunches, snarls, wild yarns, etc., can break needles of the knitting machine, and also form holes on the fabric. This shall reduce knitting efficiency as repairing the broken parts shall take a very long time. Normally these faults fall in the category of Major Six in Classimat, i.e., D4, D3, D2, C4, C3 and B4.
- **Long thin faults:** These faults if present shall prominently appear in case of T-shirts as thin horizontal lines. We get repeated bars for one or two repeats. These faults fall in the category of H and I in Classimat classification.
- **E.F and G faults:** These are long thick faults, which prominently appear on T-shirts. The "E" fault shall be very thick.
- **Contaminations:** Colour contaminations give an ugly appearance in case of bleached varieties. The HDPE or polypropylene contaminations, as they do not take any colours when dyed, appear ugly in dark coloured clothes.

3.2.2   Dyed yarns for knitting T-shirts

All the impacts discussed in Section 3.2.1 earlier for grey yarns going for knitting is equally applicable for dyed yarns also. The following are additions for dyed yarns.

- **Variation in shades:** Variation in shades creates unwanted horizontal bands, normally referred as Barre. Variation from original shade (shade specified by the customer) will lead to rejections of garments. Lot-to-lot shade variance affects garment appearance and consistency. Customers have to incur additional cost for making the fabric/garment acceptable in the market by putting some design for covering the defect due to shade variation or by salvaging the correct shade materials and group them to get the garments as per specification.
- **Colour fastness:** This is a very important requirement of any customer. Poor wash fastness leads to colour bleeding during washing. As normally different garments are washed together by ultimate customer it stains other clothes. Poor fastness to light results in fading of the garments unevenly depending on the parts exposed to sun light, giving an ugly appearance. If the T-shirts are used as sports wear, the rubbing fastness and fastness to perspiration also become critical.
- **Lot size:** The dyeing lot size should be as per the contract. If the supply is less or more, it shall cause problem in getting the exact effect that was planned. One should remember that yarns are dyed to get some special effects, and the quantity required in each shade need not be same. If any extra quantity is dyed, that yarn remains unsold.

3.2.3   Grey yarns for socks knitting

- **Count variation:** In socks knitting, the number of cones fed is less; may be one, two or four, and not very high like in the case of T-shirts where 48, 72 or 96 cones are creeled. Count variations in yarns for socks will not give problem of barre, as we get in T-shirts, as a complete sock is made up by just by a single cop or two cops. Normally a sock weighs 5–7 g against the weight of a spinning cop varying from 50 to 80 g. A variation in count results in change in the weight of socks and in some cases the dimensions of socks get changed because of shrinkage. The customer has to spend for sorting the socks as per weight/dimensions and pairing them.
- **Twist variation:** Short-term variations in twist, mainly due to higher breakages in ring frames, or uneven twist flow due to uneven yarn results in short streaky appearance. Cop to cop variations in twist,

becomes a long-term variation for socks and we might not see the effect within a sock, but we get difference in feel between socks. This also results in variations in sock's dimensions due to shrinkage and Spirality. Socks undergo more abrasion compared to T-shirts or body wears. Therefore, the yarns for socks need a slightly higher TM compared to the yarns for underwear and T-shirts; may be by 0.1. Again, the twist required depends on the end use of socks. The socks may be for baby wear, patient wear, old people wear, sports wear, casual wear, regular work wear, school children wear, etc. The baby wear, the patient wear or the old people wear needs to be soft whereas the sports wear and school children wear needs to be with higher twist.

- **Imperfections:** Short-term unevenness gives a cloudy effect, as the same yarn from one cop shall be visible side by side. The patterns due to eccentric cots at ring frame shall produce a prominent diamond like appearance on socks.
- **Strength:** Normally, strength is not a major requirement for knitting the socks, as the tension applied is very low. However, the strength required for the ultimate customer considering the type of use is important. If the socks are for sports, children wear, daily wear, etc., we need good strength in the yarn.
- **Tensile variation:** Variation in strength creates imbalance in yarn tension during knitting, and because of which we get uneven shapes in the socks.
- **Elongation:** A lower elongation gives working problem, as the yarn shall not bend properly during knitting.
- **Winding quality:** Winding defects like stitches and entanglements result in sudden variations in yarn tension while unwinding, and result in uneven shapes of socks.
- **Friction value:** A higher friction value resists the movement of yarn in the needles, and results in breakages.
- **Hairiness:** Higher hairiness results in uneven dye pick up, and sometimes in lint generation while knitting and washing.
- **Classimat long faults:** Long faults of type E, F, G, H and I, lead to unwanted thick or thin lines on socks. A periodic long thin or thick place gives a barre effect on socks.
- **Objectionable faults:** Objectionable faults like slubs, snarls, bunch, etc., lead to needle breaks, and holes formation in socks. This is very costly.
- **Consistency in cotton shade:** Although variation in cotton shade is not that critical for socks as compared to T-shirts, variation within socks is not tolerated.
- **Contaminations:** Normally, contaminations are not considered critical in case of socks.

3.2.4   Dyed yarn for knitting socks

Following additional points than that discussed in Section 3.2.3 need consideration while supplying dyed yarns for making socks compared to those discussed for grey yarns.

- *Consistency in shade:* Variation in shade creates unwanted horizontal bands. The customers do not accept variation from original (agreed) shade. Lot-to-lot shade variations affect consistency. The customers have to spend extra time and money for sorting out the socks as per the shade and make pairs so as to make the socks saleable.
- *Colour fastness:* Poor wash fastness leads to colour bleeding during washing, which also stains adjacent fabrics. Rubbing fastness and fastness to perspiration are very important for socks compared to light fastness as normally the socks are not normally exposed to direct sunlight while in use. However, if the customers are wearing skirts or half pants, the socks used are longer and covers up to calf and are exposed to sun light. In such cases, fastness to light is also important. In number of cases, although the customers are wearing full pants and the socks are not exposed to sunlight, the drying of the socks might be in sunlight. If the socks are for baby wear or for patients wear, the colour used becomes very critical as it should not be allergic or give side effects on the tender skin.
- *Lot size variations:* As few cones are fed in the creel for making socks, variations in lot size compared to order is not considered very critical. However, in case of shortages, it shall be difficult to make up the order.

3.2.5   Grey yarns for sweater knitting

- *Count variations:* Normally, sweater knitters use double or multifold yarns, and hence chances of getting higher count CV% are less. However, if the average count varies, the fabric weight shall vary significantly. The variations in fabric density change the entire look of the sweater.
- *Twist variations:* Twist variations change the fabric feel, and short-term variations result in uneven effects on the surface. The sweater yarns are normally of very low twist and hence the twist CV% are normally very high compared to that found in single yarn or with yarns of normal twist. We sometimes get completely untwisted portions. This is not accepted as the loop shall not form properly with such variations.
- *Imperfections:* Although this is not that critical as in the case of T-shirts, an increase in imperfections gives a cloudy appearance. Some customers like this appearance as it gives a woolly look.

- *Strength:* Normally sweaters are knit at a lower speed compared to T-shirts and Socks, and hence strength is not a very critical requirement.
- *Tensile variation:* Although strength is not a very critical requirement, a variation in strength introduces uneven tensions while knitting, and results in an uneven sweater.
- *Elongation:* A lower elongation gives less elasticity to the sweater, and hence looses the shape.
- *Objectionable faults:* The snarls and entanglements are highly objectionable whereas the slubs get submerged while doubling multifold yarns.
- *Winding quality:* Winding defects result in breaks and faulty fabrics. However, as the speed is very low, the defects do not affect the performance unless it is very critical like back stitches or entanglements.
- *Friction value:* A lower friction value resists the smooth flow of yarn and hence we get uneven sweater surface, and also lower knitting efficiency.
- *Hairiness:* Higher hairiness is normally considered as an advantage in sweaters, but variations in hairiness spoil the appearance of the sweater as it gives an uneven fuzzy appearance. Normally, carded yarns are used in sweater knitting excepting for some special customers.
- *Contamination:* Normally, contaminations in yarn are not considered critical in sweaters, as double or multifold yarns are used.

3.2.6   Grey yarns for underwear knitting

- *Count variations:* A deviation in average count results in variation in weight of fabric. This also affects the wear comfort and fabric feel. Variations with in and between cones give rise to barre effect, however, as the fabrics are used for underwear, this is not considered a critical.
- *Twist variations:* A higher twist gives a rough feel and affects the wear comfort. A very low twist gives linting problem while knitting and also while washing. As the undergarments are always in contact with the human body, ability to absorb perspirations and moisture is very important. Hence the twist needs to be slightly lower compared to that used for T-shirts or socks. Normally combed yarns are preferred for under garments to have better softness and absorbency.
- *Imperfections:* Increase in U% and imperfection give a cloudy or streaky appearance. However this is not critical in undergarments as in the case of T-shirts.

- **Strength:** A lower strength may result in yarn breakages, especially, while knitting at high speeds.
- **Tensile variations:** Higher variation in strength results in poor knitting performance, especially with a high-speed machine.
- **Elongation:** A low elongation results in a brittle yarn and gives rise to breakages and uneven shrinkage of fabrics.
- **Winding quality:** Uneven tension in cones results in uneven tightness factors and barre effect. Winding defects results in breaks and uneven tension while knitting.
- **Friction value:** A higher friction value resists the movement of yarn and results in breakages.
- **Hairiness:** A higher hairiness results in linting at knitting, and blocks the yarn path in the needle. This leads to higher breakages. From the point of view of appearance, this is not critical. Higher hairiness makes the fabric dirty fast. Therefore, people do not like it.
- **Objectionable faults:** The objectionable faults can break the needles in the knitting machine. This results in loss of utilisation and increase in costs.
- **Consistency in cotton shades and Micronnaire:** Normally this is not a critical issue, as in majority of the cases, the under-wears are bleached, and also the size of the fabric is normally small.
- **Contaminations:** As the fabrics are normally bleached, the colour contaminations become prominent. Although it is not visible some of the end users object this.

3.2.7   Grey yarns for knitting liner materials

- **Count variation:** Variations in average count results in higher or lower GSM of the fabric. This shall give problem in coating, as it becomes uneven. Within cone variations in count gives a barre appearance and results in cracks after coating.
- **Twist variation:** An increase in twist results in lesser absorption of the coating and a variation in twist results in cracks on the surface after coating.
- **Imperfections:** Although this is not a very critical factor for lining materials, short-term unevenness gives uneven coating on surface.
- **Strength:** The strength should be sufficient to run on the knitting machine. It is not a critical issue, unless we work on a high-speed machine. Depending on the purpose for which the liner material is used, the requirement of the strength differs.
- **Tensile variation:** Variation in tensile strength leads to uneven tensions and changes the shape of the fabric, which cannot be corrected by coating.

- *Elongation:* A variation in elongation results in uneven shape of the fabric, and shall get uneven coating.
- *Winding quality:* Uneven winding tension creates cracks in the fabric after coating. There should not be any knots as coating shall not take place properly.
- *Friction value:* High friction value creates breaks at knitting, but very high wax coating gives problem in coating.
- *Hairiness:* Higher hairiness and variations in hairiness result in uneven coating.
- *Long faults:* Long faults, either thick or thin, shall form lines on the fabric, and show as cracks after coating.
- *Objectionable faults:* These faults result in needle breaks at knitting and form holes in fabric. The coating cannot take place at holes. A slub which gets knitted absorbs more coating and results in uneven coating surface in its surroundings.
- *Consistency in cotton shade and Micronnaire:* This shall not affect the quality.
- *Contaminations:* This shall not affect the quality.

3.3   Weaving applications

3.3.1   Grey yarns for weaving apparels

- *Count variation:* A deviation in average count results in higher or lower weight of fabric. This affects the fabric feel and wear comforts. A coarser count results in lesser fabric length for the same weight of yarn, leading to a loss to the manufacturer. Variation within cones and between cones results in streakiness in warps or weft-bands in case of weft.
- *Twist variation:* A shift in twist level results in a change in fabric feel. Twist variations within and between cones lead to streaks and bars after dyeing. A softer twist results in linting and higher breakages. A very high twist results in snarling, and can entangle with adjacent ends causing breaks.
- *Imperfections:* Increase in imperfections gives a cloudy appearance in the fabric, especially after dyeing.
- *Strength:* A lower strength results in higher yarn breaks at warping and weaving. This not only reduces efficiency but also increase the hard wastes. High breakages on looms also lead to increased rejection of fabrics.
- *Tensile variations:* Variation in tensile strength results in poor weaving performance.
- *Elongation:* A lower elongation makes the yarn brittle and results in breaks. A very high elongation leads to bumping of fabric on loom.

- *Winding quality:* Winding defects result in breakages while unwinding on warping machines and in weft in shuttle less looms. Loose tension in cones results in slack ends, and hence breaks.
- *Hairiness:* A higher hairiness results in linting off in looms and also in uneven size pickup. The fabric shall have a fuzzy appearance with more hairs in the yarn.
- *Classimat long faults:* These are more critical in plain weaves. We get streaks in warp and weft due to these faults.
- *Objectionable faults:* Objectionable faults cause higher breakages on looms. The bunch ends results in multiple breaks. The snarls result in entanglement of adjacent yarns.
- *Consistency in cotton colour and Micronnaire:* Variation in shade of cotton gives variation in shade even after dyeing in certain shades and it is not accepted.
- *Contamination:* The colour contamination is not accepted as it gives an ugly look.

3.3.2 Grey yarns for heavy fabrics and industrial applications

- *Count variations*: As normally doubled and multifold yarns are used for heavy fabrics, there shall not be many problems due to count variations. Any shift in average count results in a shift in fabric weight, which shall create problems in further processes and in the strength of the fabric. A variation in count can lead to lower breaking strength because of thin portions which is critical for industrial applications.
- *Twist variations:* Variation in twist levels can reduce the strength of the fabric. As strength is the prime requirement for industrial application, we need to ensure uniform twisting.
- *Imperfections:* This is not critical for this end use.
- *Strength:* This is a very important factor for heavy fabrics for industrial applications. Lower strength of yarn shall result in lower strength of fabric.
- *Tensile variation*: A variation in tensile results in variations in fabric strength.
- *Elongation:* A lower elongation results in poor working and also a lower strength of fabric.
- *Winding quality*: Uniformity in the tension of wound packages is very essential for getting uniform fabric surface. The knot's quality is very critical. The tail end should be small and the knot should not slip. Normally, fisherman knots are insisted. In case of multifold yarns, staggered knots are recommended. Uniformity in the tension of wound packages is very essential for getting uniform fabric surface.

- *Hairiness:* This is not a critical issue for this application, excepting of higher dust liberation during weaving.
- *Classimat long faults:* These are not critical for this end use.
- *Objectionable faults:* These faults result in entanglement and multiple breaks.
- *Consistency in cotton shade and Micronnaire:* This is not critical for this end use.
- *Contaminations:* This is not critical.

3.3.3   Yarns for towels

- *Count variations*: A deviation in average count results in variation of towel weights. The variation between cones and within cones does not create problems like plain cloths for apparels as the designs and the piles cover up these variations.
- *Twist variations:* An increase in twist gives a rough feel of the towel and also shall have lesser absorbency. Sometimes we get deshaped piles because of twist variations in pile yarn. A very low twist shall result in breakages while weaving. Variation within cones or between cones is not a critical issue.
- *Imperfections*: This is not a critical issue for towels.
- *Strength*: The strength should be sufficient to work on looms without breaks.
- *Elongation*: This is not a critical issue, but should be sufficient to work on looms.
- *Winding quality*: Unwinding should be easy at warping and on looms. If the warping is done at slow speeds, the problems shall be less.
- *Friction value:* In case of dyed yarns, the finishing should be done to give a lubrication effect for the yarns to run on looms. Then yarns need not be waxed as it is done for knitting yarns.
- *Absorbency*: Absorbency is a very important criterion for the yarns for towels.
- *Consistency in shades:* When dyed yarns are being used consistency in shade becomes very critical.
- *Colour fastness:* This is a very important requirement of dyed yarns used for towels, as any poor fastness shall result in colour bleeding and spoil the appearance of the towel.

3.3.4   Yarns for carpets and furnishings

- *Count variations*: The shift in average count is more critical as it affects the weight of the fabric. The variations within cones and between cones are not considered as critical. In carpets, normally heavier counts are accepted and not fine count, as the yarn bulk is less with finer counts. The bulk of yarn is important to hold the fringes firmly.

- *Twist variations:* Normally multifold coarse yarns are used for carpet weaving. The twist per inch shall be low to get the bulky effect. As the twist is low, the problem of twist CV% is very high. A complete shift in the twist changes the effect, whereas variations with in are not considered critical. If the yarn is used for fringe, the fringes do not stand up as needed with very low twists, whereas in case of high twist portions, the fringe stands straight or bend to one direction.

- *Strength:* This is not a very important factor, although any breakages of yarn can disturb the basic design and appearance of the carpets or furnishing fabrics. As the looms run at a fairly low speed, and the yarns are of multifold in nature, normally there shall be no breaks due to weak yarn. We get breaks due to faulty winding with cones having bunches, back stitches, etc.

- *Elongation:* As a number of different fibres are used together in weaving of carpets and furnishing fabrics, the elongation variations are very high. The yarns should have sufficient elongation to work without breaks.

- *Winding quality:* Winding defects cause breakages while unwinding on looms.

- *Bulkiness*: Bulkiness of the yarns is a very important requirement of the yarns for carpets and furnishings. A low bulk makes the fabric loose and the fringes cannot be held firmly. Therefore, the level and appearance of fabric gets disturbed.

- *Imperfections*: This is not a critical issue.

- *Objectionable faults:* These faults create entanglements and breaks in weaving. Normally, carpets yarns (multifold yarns) cannot be tested on Classimat or Classifault because of their coarseness. It is a normal practice to make some hanks and physically observe the type of defects in them. The normal problems are bunches, snarls, untwisted loose ends and big slubs.

3.4   Thread applications

3.4.1   Yarns for sewing threads

- *Count variations*: This is not a critical issue as the sewing threads are used as a single thread, and not in-group as in knitting or in weaving. However, there are chances of getting a cork screw effect in the yarn.

- *Twist variations*: A higher twist is likely to create snarls while unwinding for stitching, and shall entangle to the needle. It can result in needle breakages and metal contamination in the garments. A low twist shall become soft and bulky, and shall not pass freely in the needle.

- *Imperfections:* Neps are critical for sewing threads, as they are likely to get stuck to the needle eyes.
- *Strength:* This is a very important requirement of a sewing thread, as it has to give full strength to the fabric at seems.
- *Elongation:* This is an important requirement of a sewing thread to run freely in the needles.
- *Winding quality:* Stitches on cones shall result in breaks. There should not be any loose winding which comes out or slips causing damages to the needle and the fabric being stitched.
- *Friction value:* This is an important requirement of a sewing thread.
- *Hairiness:* Higher hairiness shall restrict the movement of thread in the needles.
- *Objectionable faults:* These are highly objectionable as they break the needles.
- *Consistency in cotton shade and Micronnaire:* This is not a critical one as the yarns are used as individual threads.

3.4.2   Yarns for twines

- *Count variations:* The twines are multifold yarns and are not used for making any cloth. They are used for binding, packing, tying harnesses, etc. Hence, the variations in count of single yarn is not critical.
- *Twist variations:* The twist and twist variations is very critical as the strength depends mainly on the twist and the uniformity of the twist.
- *Imperfections:* The imperfections are not critical for twines as they are multifold yarns.
- *Strength:* This is a very critical requirement of a twine, and hence the single yarns used should have the required strength.
- *Elongation:* Elongation is also a critical requirement of a twine.
- *Winding quality:* Winding quality of single yarns is important while preparing assembly wound packages for doubling and cabling. It will not affect the quality of the twine.
- *Friction value:* This factor depends on for what purpose the twine is made. However, the friction value of single yarns become insignificant compared to the friction value requirement of final twine.
- *Hairiness:* Hairiness of single yarns is not critical, but the final twine needs to be with almost zero hairs.

- *Objectionable faults:* The twine needs to be knot free yarn. Therefore, objectionable faults including the knots are not accepted. We need to have system of staggered splicing while joining the components in the assembly winding and doubling stages.
- *Consistency in cotton shade*: This is not a critical requirement for twine.

4

Reasons for poor quality in spinning

Poor quality in spinning may be defined as a combination of factors such as the technical parameters of yarn being out of the acceptable tolerance limits of the customers, presence of objectionable faults and labels, improper packing leading to damages while handling, shabby packing resulting in time consumption while opening and extra space for storage, delay in despatch causing loss to the customer and penalty to the company, increased cost of manufacturing pushing the organization to a loss, short supplies and over productions leading to accumulation of finished material as stock. However, in practice, the terminology of quality is used for product parameters by the technicians in shop floor, but for the top management, everything is important as the company's profitability depends on the total quality and not only on the technical parameters. The technicians, therefore, are suggested considering everything together and not concentrating only on the technical parameters.

It is a normal belief that adapting latest machinery with good technology results in good quality of yarns compared to old machinery and technology. Therefore, the mills invest on latest technology. However, in spite of latest, so called "State of the Art technology" we see poor quality of yarn in a number of cases. There are number of factors responsible for the poor quality. They are raw materials, work methods, the condition of the machinery and the management systems adapted.

4.1   Raw material

There is a direct relationship between certain quality characteristics of the fibre and those of the yarn. Seventy to eighty percent of basic yarn quality is decided by the raw material. The fibre properties like the staple length, uniformity in staple length, short fibre contents, fineness, strength, elongation, maturity, etc., play a role. If the micronaire is coarse, the number of fibres in the yarn cross section will be less. This always results in lower strength and lower elongation. But it is easy to process coarse micronaire fibres in blow room and cards. Nepping tendency is less for coarse micronaire fibres. On the contrary, spinnability (in both speed frame and ring frame) is not good with coarser micronaire fibres. U% is affected by micronaire. Coarser the micronaire, higher shall be the U%. Coarser the fibre, higher the end breakage rate in spinning. Uster thin place (30%) in the yarn vary depending upon the fibre micronaire. Lower the micronaire lower the thin places and vice versa.

Raw materials if not suitable for the product being produced cannot give the quality parameters as needed. We need to understand the quality objectives of the product and decide the raw materials. The factors like feel of the fabric, hairiness in the surface of fabric, strength of the yarn and the fabric and dyeability are governed by the raw material properties. Just by spending more money on superior fine cotton cannot give the quality required by the customer, but shall increase the cost of manufacture leading to a loss to the company.

Raw material should be suitable for the machine being worked. The machines are normally designed for a range of raw materials, and we need certain modifications while working. The type of opening machines differ from cotton to synthetics, and also with in the cottons itself from fine to superfine cottons. Coarser wire points are used for carding polyester, whereas for cottons we have different wire configuration. The nose plates are different for cotton and polyester. In ring frames, large diameter bottom rollers of 30 mm are used for long staple synthetic fibres compared to 27 or 25 mm diameter for cottons. The cradle sizes used in drafting are different for different staple lengths. Similarly there are number of changes in the machine configuration to suit different raw materials. If we are not clear about the same, inspite of having latest machinery and good raw materials, we might produce bad quality yarns.

Variations, mix-ups, contaminations and damages in raw materials are also responsible for poor quality yarns.

The selection of raw materials for the product and the machines play a vital role in getting the required quality. Apart from the selection of raw materials, the ways in which they are handled and used also play an important role. Following are some examples.

4.1.1   Improper selection of raw material

Let us take some examples of improper selection of raw materials for a given end use.

(a)   Superfine cottons/fibres where coarse cottons/fibres was a need

    i.   Customer wanted carded Ne 20^s for making T-shirts on an urgent basis. The mills were running Ne 40^s Combed using fine cottons. Instead of taking a separate mixing, the mills decided to give Ne 20^s K with the running mixing only. They produced the yarn and the quality of yarn for Ne 20^s K was very good from the point of view of the spinner. The U% was lower, the hairiness was very low, the strength was good, the imperfections were low and the elongation was good and so on. The management of spinning unit was happy although the cost of manufacturing was high due to use of fine cottons, with an anticipation of customer coming back to them again with bigger orders. However, after a period of three months, the spinner received a feedback from the customer that the T-shirts were not accepted in the market as it had a "feminine feel" and the end customers wanted a "masculine feel".

ii.   The customer wanted Ne 7^s K for weaving rough jeans. The spinner used cottons having 3.8–4.0 micronaire. The Jeans had a soft feel and was not accepted by the young customers. They wanted a rough cloth, and hence a micronaire of 4.7–5.0 should have been good.

iii. A customer wanted to manufacture towels and a rough surface was the requirement. The count of the yarn required was Ne 2/20^s. The spinner used superfine cottons which gave a smooth appearance. The customers did not like the feel.

(b) Coarse cottons/fibres where fine cotton/fibres was a need

i.   It is necessary to use fine cottons to spin yarns of fine counts. However, it is observed a number of times that due to price considerations, the mills over-spin the cotton resulting in poor productivity and quality.

ii.   A customer ordered for Ne 10^s combed for the purpose of embroidery, whereas the spinner was not aware of the end use. He used normal coarse cotton for the same. The customer was not happy because of the hairiness.

iii. A customer wanted Ne 20^s K for weft in a tyre cord. There was a requirement of minimum 600 g strength, (20 RKM) whereas the spinner was not clear about the same. He used normal medium cottons and gave an RKM of 16–17, which was not accepted by the customer.

(c) Giving weightage for parameters without understanding their impact on the quality requirement

Numbers of mills have a practice of testing cottons from each bale and grouping the bales as per the quality. The mills spinning fine yarns for bleached varieties need to group the bales as per length, whereas for the mills spinning yarns going for fabric dyeing it is preferred to group the bales as per micronaire and maturity. In case the yarns are required for industrial applications, the strength becomes a major factor. In a number of cases it is seen that the mills purchase readymade software for bale management and go by the default fed in the software, without understanding the basic logics of grouping.

(d) Not verifying the compatibility of components in a mixing
The fibre properties and their proportion in a blend or in mixing is very important. The component fibres must be compatible to each other. If fibre properties are not compatible in terms of length, fineness, frictional properties and elongation, it is likely to give trouble in processing due to possibility of de-blending. Differences in fineness affect the yarn quality adversely as it affects drafting behaviour, migratory behaviour, dye pickup and differential speed of fibres during pneumatic transport. This can give rise to strength related problems, streakiness and barre in the dyed fabrics. A large difference in elongation of fibres between the components leads to problems in drafting and also adversely affects the yarn strength due to unequal sharing of tensile load.

4.1.2   Variations in raw material

(a) We normally see variation in length, strength, elongation, maturity, colour, fineness from lot to lot and bale to bale. The variations may due to any reason like natural variations in soil conditions, seasons, seed quality, work methods, etc., and also due to purposeful act of mixing up some poor quality materials to get higher price realization.

(b) Mix up of two varieties of cotton takes place either by ignorance or by poor working methods at ginning area.

(c) Mix up of two merge numbers in case of synthetic staple fibres results in shade variations after dyeing, and sometimes working problem. This mistake normally takes place in the raw material godowns of a spinning mill because of ignorance of workmen or issue clerk.

(d) Season wise variation in raw material properties is seen especially in micronaire values, colour and strength.

4.1.3   Handling and storage damages

(a) Fungus attack takes place due to storage of bales in a damp area, storing bales on ground in fields, applying more water while pressing the bales, etc. This reduces the strength of fibres and also gives problems in dyeing.

(b) Contaminations of unwanted materials like Hessian twines, polypropylene, coloured cotton thread or cloth pieces, polyester cloth pieces, etc., is one of the major problems faced by the spinners. The effect of different contaminants on the yarn quality and process performance can be summarized as follows:
    i.   Strings and fabrics of jute lead to increased breakages in rotor and ring spinning, poor appearance and differential dye pick-up.

ii.   Strings and fabrics of dyed cottons lead to poor quality of yarn and cloth due to coloured fibres.

iii.   Strings and fabrics of woven plastics and plastic films lead to differential dye pick-up, very poor yarn and fabric quality and also damage to machinery.

iv.   Organic matters like leaves, feathers, papers, leather, etc., lead to damages to machinery and increased wastes at spinning.

v.   Inorganic matters like sand dust, metals particles, etc., lead to damages to machinery, increased wastes at spinning and to fire accidents.

vi.   Oily substances and stamp colours mar the yarn and fabric appearances.

vii.   Human hair contamination leads to increased end breakages at ring and rotor spinning, poor yarn and fabric appearance and also to differential dye pick-up.

viii.   Presence of stones in cotton damages the machinery and also leads to fire accidents.

ix.   Seed coats result in poor appearance of yarn and fabric and more wastes at spinning.

x.   Materials like pouches of Gutka lead to machinery damages and poor yarn appearance.

(c)   Fibre damages in the surface of bales because of dragging. This problem is seen while loading and unloading trucks which have old wooden floor.

4.1.4   Damages due to improper ginning in case of cotton

Improper ginning results in higher neps, half broken seed bits, short fibres, etc. Half broken seed bits are very difficult to remove and results in loading of cylinders and flats surfaces, blocks the screens in under casings and grids resulting in poor cleaning of cottons. Poor ginning results in higher neps in yarn, oil stains due to broken seeds, and uneven yarn and lower strength due to fibre rupture.

4.1.5   Fused fibres in case of synthetic staple fibres

The fused fibres are hard particles present in synthetic fibres due to some problems in heat setting area. They result in loading on card surfaces leading to neps formation.

4.1.6   Presence of uncut fibres in staple fibres

This problem is mainly due to failure of cutters in fibre spinning plant. It is very difficult to identify this problem while inspecting the bales or testing them, as these uncut fibres are distributed randomly in a bale.

4.1.7   Packing of cotton bales with polyester cotton cloths/HDPE/polypropylene cloths

It was a normal practice to pack the cotton bales with Hessian for a long. During 1980s as jute became very costly, the ginners started using woven HDPE cloths for covering cotton bales. As the problem of contaminations was prominent with both Jute and HDPE coverings, the mill people started insisting on covering the cotton bales by pure cotton cloth in 1990s. It went on well for some years. As the polyester became cheaper than cotton, the ginners started using polyester cotton blended fabrics for covering cotton bales, which was looking more like cotton only. This started a new problem of polyester fibre contamination, which is more difficult to remove even by using contamination detection units like Optiscan or SIRO clearers.

4.1.8   Inadequate number of lots in a mixing

The number of lots allotted for a mixing, sometimes, might be inadequate compared to the plan of one bale from one lot. This type of problems are normally due to short supply of cotton in market, financial crisis because of which the mill is not in a position to procure more cotton and at the end of a season where the old cotton lots are being run out. Sometimes, when the mills are doing job works, the customer does not send more lots of cotton to reduce the interest on working capital. When more bales are taken from any lot, that lot shall run out faster and we need to replace it with another lot with the same quality parameters, which is difficult. Hence this leads to variations in yarn properties.

4.2   Work methods

The work methods are very critical for getting the required quality of yarn. One can get a good result inspite of fairly poor raw material or old machine provided appropriate work methods are followed. We can group the work method related factors as follows:

(a) **Old systems/practices not modified as per the new requirements:**
The technology is continuously changing and also the requirements of the customers. The work methods are developed by considering the technology and the facilities available, the culture and the customer

needs at the time of evolving a new work method. As the technology and the facilities change, we need to modify the work methods.

(b) Inadequate training and lack of knowledge: The skill levels and the knowledge of people for the job and technology are very important for getting the best results out of the available materials and machinery. Inadequate knowledge and lack of training can spoil the materials and machinery. The higher employee turn-over and the absenteeism in spinning mills add to this problem. With the increase in literacy level and emerging of new job avenues in electronics, information technology, tourism, etc., the mills are finding difficult to get workmen.

Let us discuss the normal bad work practices seen at different processes of a spinning mill.

4.2.1   Mixing

1. **Not verifying the lot numbers and bale numbers before taking the mixing:** The raw material lot properties are studied before deciding the components of a mixing and quantity to be taken from each lot is decided by the top management. It is the duty of shop floor technicians to verify whether the mixing issued is exactly the same as per the plan given by the top or not. There are chances of getting bales from different lots because of miscommunications, confusions and illegible lot numbers and bale numbers on the bales.

2. **Taking more bales from a lot for mixing:** It is always suggested taking cotton from more number of lots for a mixing rather than from just two or three lots. By taking small quantity from each lot, we can get uniformity in mixing characteristics for a longer time. Instead of taking one bale from each lot, if we take more bales from a lot, the lots run out very fast. This gives fluctuations in yarn properties.

3. **Not clearly defining/understanding the quantity of materials to be taken from each mixing component:** It is the duty of the top management to specify clearly the quantities of each component in a mixing and not to leave it on juniors. Also they need to supervise whether the mixing components are taken as per their plan. The same should be cross verified by periodic physical stock taking of bales from different lots and tally with the consumption figures shown.

4. **Uncontrolled mixing of soft wastes:** In order to improve the realization and reduce the cost, the reusable soft wastes are added back in the mixing, either in the same mixing or in the next lower mixing. However, there is a need to monitor the quantity of wastes added in a mixing. A higher percentage of soft wastes added results in higher

neps, uneven yarns, lower strength and poor working. It is seen in numbers of mills that the soft wastes generated are put in the mixing back along with the running mixing in the shifts. This is a very bad practice as there shall be no account and the quality of mixing does not remain same. The mixing processed prior to this shall have less percent of soft wastes and the percentage increase very high when the waste is added. The soft wastes should be added only at the time of preparing the mixing and should never be added once the mixing is working in the blow room. One should ensure that the soft wastes are mixed homogenously at all places of the mixing.

Workers or supervisors in the shifts mix the soft wastes in the mixing without proper accounting to safe guard themselves from their bosses for excess soft wastes produced. If the seniors work with the supervisors and workers, analyse the reasons for excess waste generation and take suitable actions, this type of mistakes can be reduced. Only firing the supervisors cannot improve the situation.

5. **Putting all soft wastes at one place in a stack:** It is seen in a number of mills, that soft wastes are brought in bags and just unloaded on a mixing by the production workers. The mixing coolies do not bother for opening those soft wastes and spreading them uniformly. The soft wastes should be opened thoroughly and spread all over the mixing area while mixing is being prepared, and not to be put on an already prepared mixing. In number of cases we can see the sliver wastes and roving wastes in the mixing bin with long lengths. These can damage the lattices of hopper feeders and bale breakers. Therefore, the workers are to be educated properly and their activities monitored by the supervisor.

6. **Not taking materials from each bale uniformly while making a stack mixing:** In stack mixing, normally 15–20 bales are taken and five to eight workers shall be working. Each worker would have been allotted certain bales to be opened. One worker might be faster than the other, or might not be opening the tufts thoroughly. Because of this the workers might complete their bales first and put them in the mixing, whereas others are still lagging behind. By this there cannot be homogenous mixing. Cottons of some bales shall be concentrated at one place whereas at other places there shall be no representation of this bale.

7. **Not keeping the bales in allocated place while mixing or feeding to bale pluckers:** It is always suggested to demark place for each lot/bale while keeping the bales. This shall help us identifying the lot and bales in case of any problems of poor quality or contaminations. Also this shall help in ensuring that the mixing is issued as per the plan and there are no misses. In good mills a display

board shall be put showing the lot numbers and bales that are working and their positions.

8. **Not removing the Hessian/HDPE covering and bale hoops properly:** It is very essential to remove the Hessian/HDPE covering and bale hoops from the mixing area and deposit at a designated place immediately after each bale is opened or else it can result in contaminations and accidents. The bale hoops are to be folded immediately after they are taken out from bales.

9. **Not opening the cottons thoroughly:** The workers are supposed to open the cotton by hand thoroughly before putting in a mixing. However, it is seen that they are unable to do it mainly because of the workloads allotted to them. They put big lumps in the mixing. It is also seen that even if the work is less, the big a tufts are put in mixing as a human tendency. However, this problem is not there where ladies are employed for checking the contaminations from cotton bales as just one or two bales are allotted for a lady compared to 15 bales allotted for a male mixing coolie.

10. **Not removing the big visible contamination while opening the bales and doing mixing:** Anyone who sees a big contamination in a bale or in a mixing is supposed to remove it. However it is seen that people are busy in their works and do not bother to remove it immediately once it is seen. This is just a human nature.

11. **Keeping chappals/slippers, etc., near the mixing bin:** In many mills, workers do not have the habit or culture of wearing shoes or chappals while they are on work. It is seen that in a number of cases proper place in not allotted for the workers to leave their foot wears. They keep their chappals either near the machines where they are working or near the mixing bin. Sometimes, the cottons fall on them while taking from bins to the bale breaker, and then get into the mixing. This results in heavy damages to the blow room machinery.

12. **Not cleaning the floor thoroughly before laying a new mixing:** It is very essential to clean the floor thoroughly before starting a new mixing to avoid unwanted contaminations.

13. **Not updating the identification boards:** The identification boards indicating the mixing, the date and time of mixing done and the time after which the mixing can be used in blow room are to be updated immediately after preparing a mixing. Sometimes, these boards are not updated leading to confusions and improper conditioning

14. **Pouring more water/cotton spray oil at one place:** In dry areas, there is a practice of spraying cotton spray oil and water on the mixing. In such cases, care should be taken to use a fine spray by using spray pumps or spray guns. Although good spray guns or pumps with good nozzles are used, it is seen that at the end, when the quantity of water

in the bucket becomes very less, the bucket is emptied by pouring the remaining water at one place on the mixing, by which more water or cotton spray oil falls at one place, and it results in entanglement of cottons.

15. **Handling the mixing in wet condition:** The mixing should not be touched till it is dry after spraying the cotton spray oil or any antistatic. If the material is handled in wet condition, the cottons get entangled and matted.

16. **Sleeping on the mixing bin during leisure hours:** It is seen in some mills, the workers sleep on the mixings during their leisure time, especially during lunch hours. It is a bad practice and should be discouraged. Management should provide alternate place for the workers to take rest.

4.2.2   Blow room

1. **Improper selection of cleaning and opening points for the mixing in hand:** The selection of opening and cleaning points depend on a number of factors like the trash content in the bale, micronaire value and the density of the bale. The cottons with higher trash need more cleaning, and hence more beating points are used. However, if the micronaire is low, then it is not advisable to have more beating to clean the cotton although the trash is more. We need more opening points in such cases. Coarse cleaning equipments are designed to extract heavy, large, loose trash particles like husk, stem, stone chips, sand, leaf particles, etc., and cannot be made effective for removing finer trash particles and seed coats. Fine or intensive cleaning equipments are designed to be more effective on finer trash particles.

2. **Not cutting the mixing vertically while feeding to mixing bale opener:** While taking cottons from a mixing stack, it is necessary to cut the stack vertically as the cottons from different bales are laid horizontally. However, when the heights of the stacks are very high, the workers find it difficult to cut the complete height. They collect materials from top three to four layers first and again collect from the bottom layers. It is therefore suggested to have lesser heights of the stacks. Where space is a problem and more area cannot be allocated for stack mixing, it is suggested to have additional mixers in the line like multi-mixers or auto mixers.

3. **Using improperly conditioned mixing in case of stack mixing:** The stack mixing needs conditioning for minimum 24 h. If the conditioning is not proper, the moisture content in cotton shall not be uniform, which might lead to uneven weights of the laps, uneven opening, etc.

4. **Feeding big lumps of cotton to bale breaker:** Where the hand

mixing is done, it is seen that very big lumps of cotton are fed to the feed lattice of the bale breaker. The big lumps need sever beating to get opened, and also are likely to damage the spikes in the inclined spike lattice and evener rollers. If the tuft size is small, a gentle opening can clean the cottons without damaging the fibres.

5. **Widening the settings to have more production:** In a number of mills it is seen that the settings are widened between inclined spike lattice and the evener rollers in the bale breakers so that the production could increase. Similarly the settings in hopper feeders are also increased. This might give the production as needed, but results in improper opening and cleaning. The load on carding increases if the blow room has not done its work properly. Increase in load on carding results in fast wearing out of card clothing, which is very costly.

6. **Too close a setting resulting in over beating and fibre damages:** In order to get good opening and cleaning, sometimes, the settings are made close. Too close a setting leads to fibre damages. Hence it is necessary to study the fibre rupture percentage before deciding on a close setting.

7. **Laps of longer lengths:** In order to increase the working efficiency of blow room as well as cards, it is suggested to take the maximum advantage of higher lap length. However, there is a limit for the same. Very high lap lengths result in higher diameter of lap which is difficult to handle. As the lap length increases, the lap weight also increases. This results in increased pressure on the inner layers of the lap as the lap is building up. It finally results in lap licking at cards. It is therefore necessary to study the practicability before deciding on the length and weight of laps.

8. **Applying excessive weights on lap spindles to get a compact lap:** Laps with higher diameter is difficult to handle, and hence people prefer the laps to be compact. They increase pressure on lap spindle for this purpose, which is a wrong practice. Increasing pressure on lap spindle leads to the problem of lap licking. It is therefore suggested to have more pressure on calendar rollers and make the lap sheet compact rather than trying to apply more weight on the lap spindle.

9. **Not removing the accumulated wastes in time resulting in back firing and waste contaminations:** The droppings collected below the beaters are to be removed in time and not to be allowed to accumulate. If the wastes are accumulated, there are chances of them getting sucked back resulting in trashy material concentration at some places. Where the suction systems are made for removing the wastes from the beaters, we need to ensure that the paths are not blocked by the wastes. If the paths are blocked, the suction shall be poor and the air currents take a reverse direction leading to back firing (see Fig. 4.1).

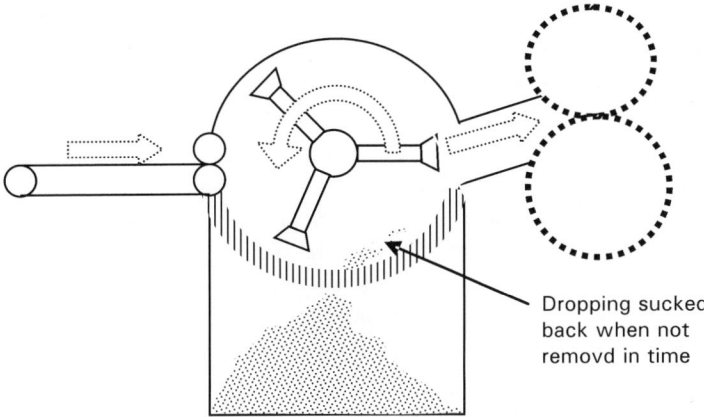

4.1 Blow room process.

10. **Not cleaning the magnets on time:** In blow room lines magnets are provided at a number of places to catch any metal particles present in cottons fed. When a metal is caught by a magnet, it obstructs the movement of other cottons and hence there shall be an accumulation. This accumulated cotton and the metal piece should be taken out immediately. To facilitate taking out of the metal, normally windows are provided by the side. If the accumulated cotton and the metal pieces are mot taken out, the accumulated cottons carry the metal forward because of the suction pressure in the ducts. Therefore, the purpose of installing a magnet is not served. These metal pieces can damage the machine parts and also can result in accidents.

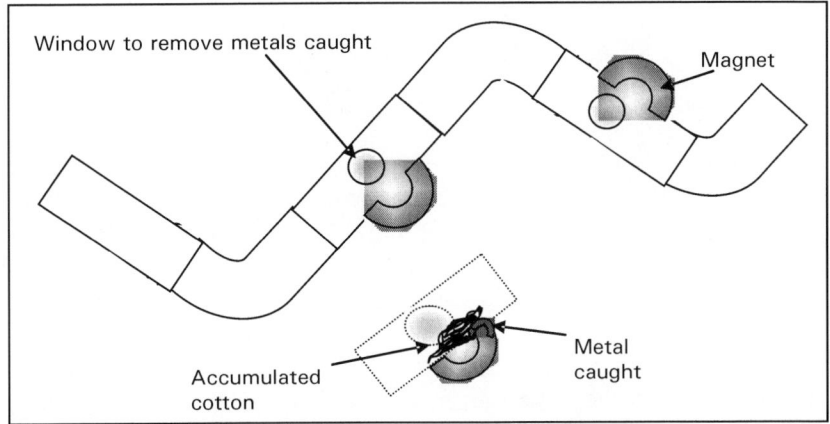

4.2 Blow room lines.

11. **Not properly closing the doors:** Almost all machines in blow room are provided with side doors for various purposes like safety, aesthetic

look, etc. However, it is seen in number of cases that the doors are either kept open or not fully closed while working the machine allowing air leakages, dust contaminations and also leading to accidents. There are mills where these machines also serve as a storing room for machine parts, records of the company and personal belongings of the employees. It is a very bad culture as it can affect the quality of the materials being processed, and also can be harmful to the employees. In case of fires, for which blow room is notoriously known, the material kept inside the machines can be lost. Sometimes, these materials shall be the root cause for accidents and fires. Some blow room manufacturers have provided a place inside the machine to keep certain spares like change gear wheels, sprocket wheels, etc., and have provided slots for them. We should keep only those materials inside the machine. But closing the door properly is a must.

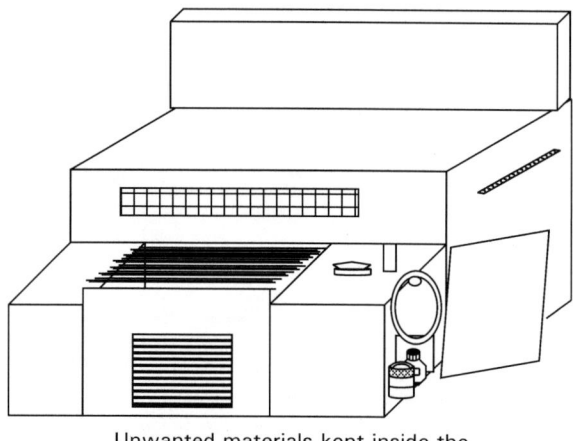

Unwanted materials kept inside the machines and on the machines

4.3 Unwanted materials kept inside and on the machine

12. **Not properly following colour codification for laps:** In the earlier days chalk marks were being used to identify the mixings of a lap. As the awareness improved and it was found that chalk marks were leading to colour contaminations, the practice of using a coloured disk on the lap spindles were started. The disks may be either of a metal or of plastic. It is seen that these discs are not properly maintained and sometimes fall off resulting in a confusion as to which mixing the belonged to. Hence chances of mix up are there.

13. **Not maintaining the work area clean:** If the work area is not maintained clean, there are chances of contaminations. A clean work area helps in identifying the mistakes faster than a dirty work area. This is the reason, it has been considered as one of the important

concepts in 5-S. If the machines are maintained clean, any cracks or loose parts can be seen easily. The accumulation of fluff on the walls, panels, roof, corners, etc., are common in a blow room if the doors, openings and joints not sealed properly. The accumulation of fluff helps spreading of fires in case of sparks and fire accidents. If the section is clean, it prevents the fire accidents, and a stray spark shall just vanish out instead of getting converted into a fire.

14. **Not cleaning the sensors in photo electric controls:** The photo electric controls sense the level of cotton by the shadow fallen on the sensor. In a number of cases it is seen that the dust get accumulate on the glasses and the sensors, because of which the sensors shall be sensing the presence of cotton. This results in improper feed to the next machines.

15. **Using of bent lap rods:** The lap rods become bent as the laps are kept as reserve in a carding machine. The lap rod is supported by two ends whereas the weight of lap is acting in the center resulting in a bent lap rod. A bent lap rod can lead to excess stretch while the laps are running out on cards although the lap feed is by the surface contact of lap roller.

Spare lap kept on card creel
Weight of lap is acting on the lap rod

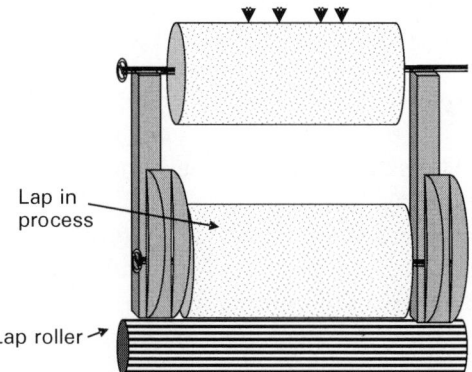

Lap in process

Lap roller

4.4 Bent lap rods.

16. **Using torn filter bags:** The purpose of filter bags is to filter the dusty air and allow the filtered clean air in the departments. The filter bags are made of canvas cloth with fine perforations. The bags are always under pressure because of the blowing of air. Therefore, there is always a chance of bags getting burst or torn out in order to allow the air out. When a torn bag is used, all the dust will enter the production unit and settle on clean materials making them dirty. This is also injurious to the people working in the production area.

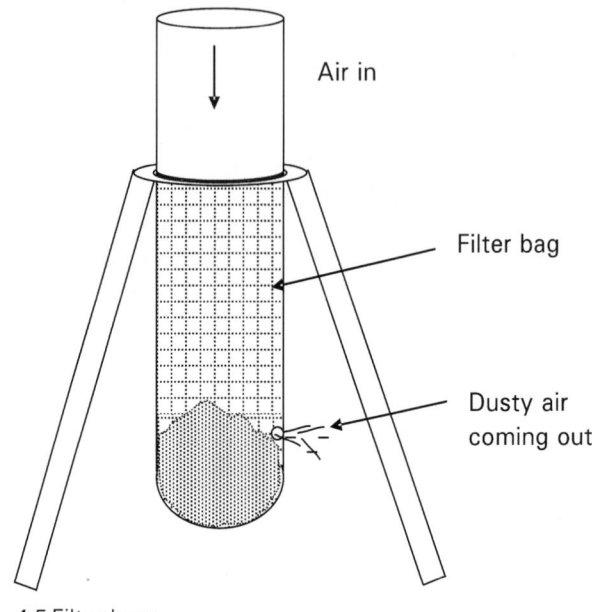

4.5 Filter bags.

17. **Cleaning the machines with brooms while working:** In number of cases it is seen that the workers clean the sides of blow room with a broom while the machines are working. In this process, sometimes, they touch the running parts like belts, pulleys, etc., resulting in accidents and contamination of broom sticks in the cotton.

18. **Using compressed air for cleaning while machines are working:** Compressed air is required for blow room for actuating the pneumatic valves and pistons in the by pass arrangements, feed regulators, two-way distributors, and in lap doffing units. If the workers are allowed to use the compressed air for cleaning the machine parts while the machines are working, it results in inadequate supply of air for pneumatic controls. This might lead to malfunctioning of valves leading to improper feed and poor quality.

4.2.3   Carding

1. **Bringing the laps on head or by touching it to the body:** In some mills, the workers have practice of carrying laps over their heads or holding them by their hands touching to their chest. Both are bad practice, as it can lead to human hair contamination and also oil contamination. Workers should make use of trolleys specially designed for carrying laps (see Fig. 4.6).

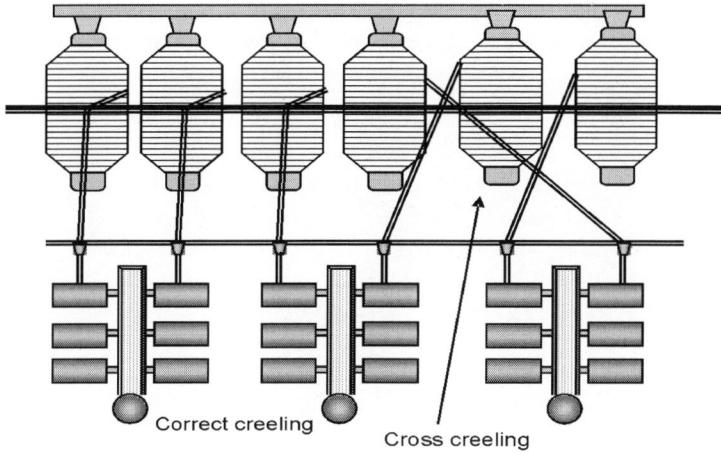

Correct creeling   Cross creeling

4.6 Trolleys designed for carrying laps.

2. **Not properly feeding the laps resulting in doubles or singles in sliver:** When a lap is being run out, the worker brings a new lap and keeps it as reserve. He removes the last half metre of lap as it shall be normally folded and uneven. Then he feeds the sheet of fresh lap manually by cutting and adjusting the bits. Care should be taken to avoid overlapping, that can result in doubles and voids that can result in singles. The skill of the worker plays an important role.

3. **Not removing the doubles and singles:** Due to various reasons like folds while feeding laps, torn laps, lap licking, partial breakage of web after doffer, we get either doubles or singles. It goes unnoticed a number of times. But when anyone sees it, it is their duty to stop it. Ignoring a single or double and working the machine just to get production is more harmful than not producing the material.

4. **Not removing the wastes in time:** Wastes from below the card needs to be evacuated on a regular basis. If the wastes are not removed in time, they might be sucked back resulting in heavy damages to the machinery and materials (see Fig. 4.7).

5. **Not keeping the carding area clean:** Keeping the carding area clean is very important to avoid stains, fluff contaminations and also potential fire accidents. If the allies are not clean, it might also result in accidents to workers.

6. **Not cleaning the card sides with a clean broom stick regularly:** It is normally found that sometimes wastes shall be hanging at the sides of the cards resulting in torn selvedges of the web, slubs and holes in the web. Good tenter always ensures that the card sides are clean. As the cylinder, doffer, licker-in are revolving and have sharp wire points covering their surface, one should not try to put his hand for cleaning.

Also no brooms or compressed air are to be used while cleaning the sides. The safest method is to use a thin broom stick, and rotate it by hand to remove the accumulated fluff. Some companies have developed fluff gun to remove the fluff. However, as the thickness of fluff gun is more and is not easy to handle compared to a broom stick, the workers prefer using a broom stick rather than a fluff gun.

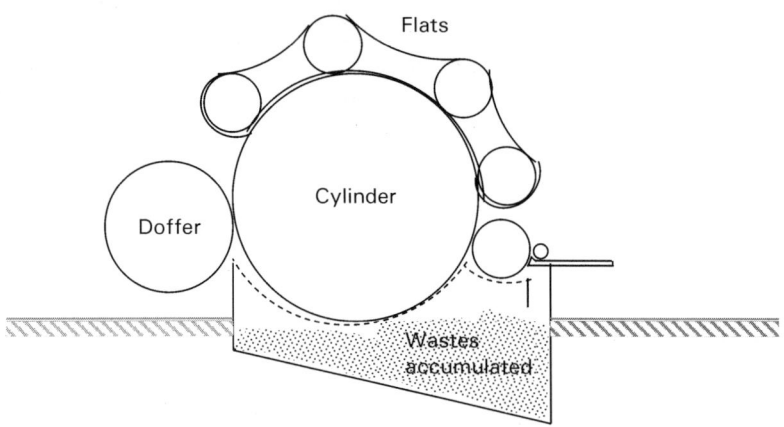

4.7 Wastes accumulated at sides of cards.

7. **Twisting too much while piecing the slivers:** The piecing of slivers is done by hand. In the end, the jointed portion is slightly pressed and twisted to avoid them getting separated. If the pieced portion is compressed much than needed, it shall be difficult to open it in the next process.

8. **Not following the correct colour codification and channeling:** Colour codification is done to ensure the right material being worked. Non-following up of colour codification leads to mix up of materials from different mixings and blends.

9. **Not stripping the cards in time:** In cards with flexible clothing, the cylinder and doffer gets accumulated with cottons and trash. It is needed to strip them periodically. In case of modern cards with metallic clothing, the loading of wire points are rare, however, if the feeding is heavy, we get some loading problem. In such cases, it is needed to strip the card and restart. If we do not strip the cards in time, we end up with neppy sliver. The card, instead of removing neps, shall generate neps and spoil the yarn quality.

10. **Not removing the accumulated flat strips in time:** The flat strips need to be removed periodically. If it is not removed, it might touch the flats and get back on to the flats. Sometimes the wastes are sucked by the cylinder from the small gap between front plate and the

stationary plate. This might result in damages to wire points and hence the quality of web produced (see Fig. 4.8).

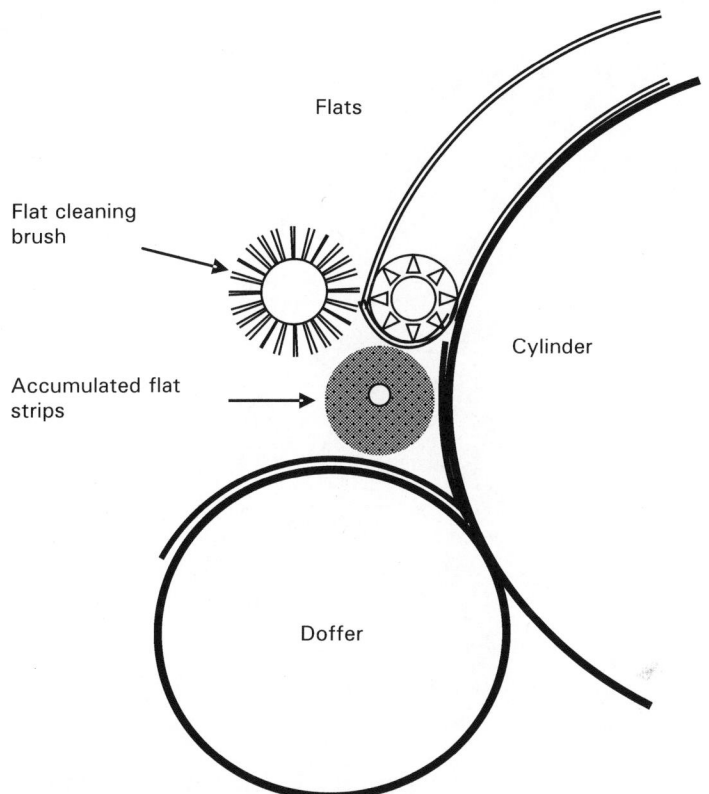

4.8 Wastes accumulated at flat strips.

11. **Not cleaning the Phillipson rollers in time:** The revolving flat cleaning brush is also known as Phillipson roller. It should be cleaned periodically. If it is not cleaned, the flats take back the accumulated trash materials from the brush, and this gets contaminated with good material (Fig. 4.9).

12. **Making the stop motions inactive:** Stop motions are provided for the safety of the persons working and for the safety of the machines. It is seen in number of cases the operatives inactivate the stop motions either purposefully or by ignorance. The stop motions, sometimes, due to loose connections or improper setting, might act; because of which the productions may slightly come down. In such cases, the workers inactivate the stop motion instead of referring to the supervisor and getting it rectified. In some cases, while attending a break down or while cleaning certain parts, the stop motions may be

made ineffective by the maintenance person, but is not restarted while starting the machine.

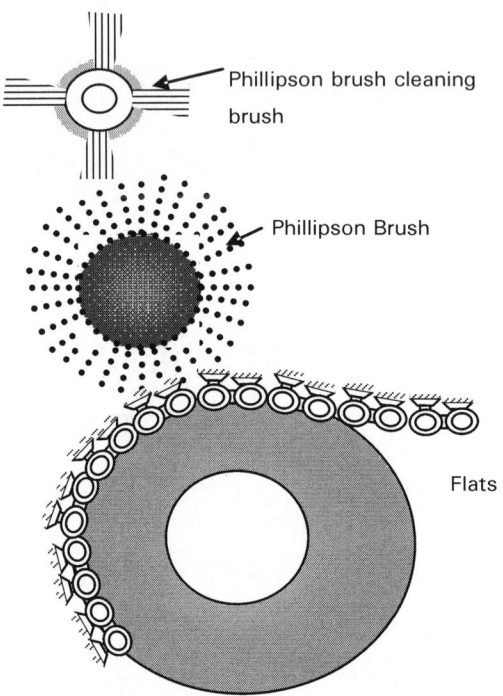

Phillipson brush cleaning brush

Phillipson Brush

Flats

4.9 Phillipson brush.

13. **Keeping the doffer cover up while running:** The doffer cover is having two functions; to protect the doffer and the person working, and also to remove the loose floating fibres. The doffer covers are normally is fitted with a suction devise to remove floating fibres. However, it is seen in numbers of mills that the covers are not put back by the workers after attending a break. Because of this the floating fibres are not extracted and sent out, and the working area becomes dirty. The fluff gets contaminated with clean good materials, resulting in various types of faults in the yarn (Fig. 4.10).

14. **Keeping the side covers and belt guards open:** The side covers and belt guards are provided for the safety of workers and also of the machine. Hence they should be always kept closed. Safety of employees is one of the main aspects of quality, as only the skilled and experienced workers can give the quality. Skilled employees are the real asset of any company.

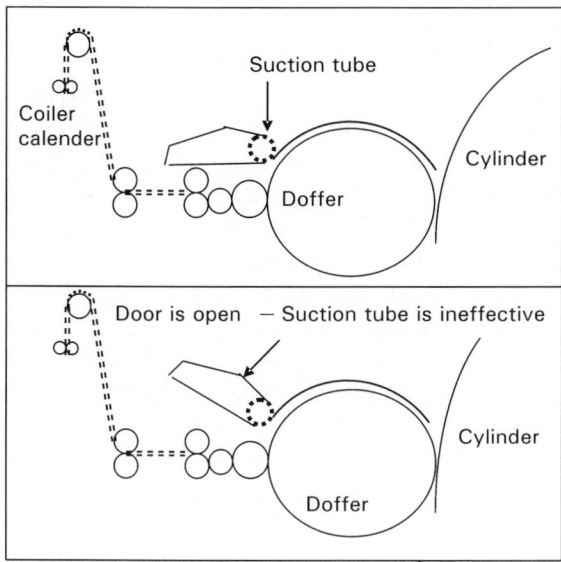

4.10 Side covers and belt guards.

15. **Over filling the cans:** In a number of cases, it is seen that the cans are overfilled; may be because of shortage of cans, or not bringing and keeping the cans near the machine in time. To avoid stoppages at draw frames and carding the workers allow the cans to overfill. One should understand that overfilling spoils the quality of sliver as the layers get rubbed with the revolving coiler plate. If the slivers get compressed, they stick to adjacent coils and licking shall take place while taking the slivers out (Fig. 4.11).

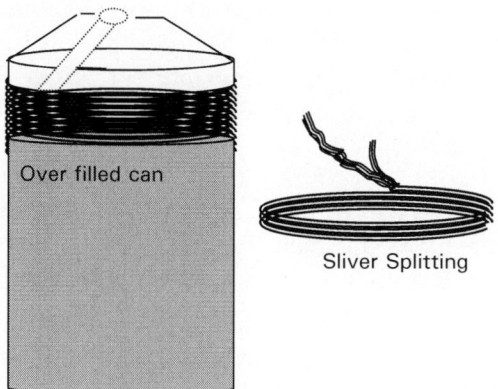

4.11 Overfilling the can

16. **Not removing the wastes from can bottom before placing below coiler:** The wastes from the can bottom is to be removed before

putting the cans for filling. But in a number of cases it is seen that the system is not followed religiously. This leads to contaminations and mix ups.

17. **Using damaged cans and spring plates:** The damaged cans and springs disturb the slivers and splitting takes place. Therefore, damaged or dented cans and springs should not be used. All damaged cans are to be removed out of the work place, as the workers might use as the cans are available in the work area. The management should ensure that all broken cans are promptly removed and discarded.

18. **Not cleaning the castors regularly:** The castors are provided for the free movement of cans. However, if the castors are not cleaned and maintained, it gives jerks while transporting the cans disturbing the coils. This might lead to higher sliver wastes in next process. Also, it can result in stretch in sliver leading to count variations.

19. **Pressing the sliver in the can by hand:** In order to fill more material or to asses the quantity of sliver in cans, the workers, and sometimes even the supervisors, press the sliver in the can. It should be understood that the pressing of this type disturbs the slivers and coils and might result in wastes and count variations because of stretch and splitting.

20. **Not covering the sliver in cans that are in stock:** The slivers after doffing from cards are to be covered to prevent fluff falling on them. If the machineries are in a good condition and suctions are working effectively ensuring a clean atmosphere, the covering of cans need not be done. It is more essential in old mills, where proper humidity controls with adequate air changes and micro-dust extractions are not there.

21. **Use of compressed air for cleaning while machine is working:** In a number of mills it is seen that the people including senior technicians and managers use compressed air to clean themselves and the machines. The compressed air is provided to activate the pneumatic controls in the machine. The irregular and uncontrolled use of compressed air results in malfunctioning of controls. This lead to a bad quality. Also while cleaning, lot of dirt flies up and settles on other machines and materials already produced, resulting in contaminations. It is not recommended to use compresses air for cleaning self, as it is injurious.

4.2.4   Draw frames

1. **Not creeling the cans as per the allocation given (Card number wise or drawing delivery wise):** It is essential to maintain the can position in the draw frame creel according to the card. This shall help

in identifying the rouge cards and correcting them, and also ensure that materials from all cards are blended homogenously leading to consistency in quality. It is suggested to number all cans with the card numbers and use only those cans for the respective carding machines. The card number might be displayed on the creel of the draw frame.

2. **Keeping the feeding cans much away from the creel:** The feeding cans should be kept as near as possible to the creel to avoid stretch while taking the slivers. This is very important as the sliver does not have any twist and can be stretched easily. It should be noted that a small stretch in draw frame sliver is multiplied hundreds of times by the time it is converted into yarn, and results in count variation and strength variations.

The effect of stretch is more incase of combed sliver, as the fibres shall be highly parallel in the sliver compared to card slivers.

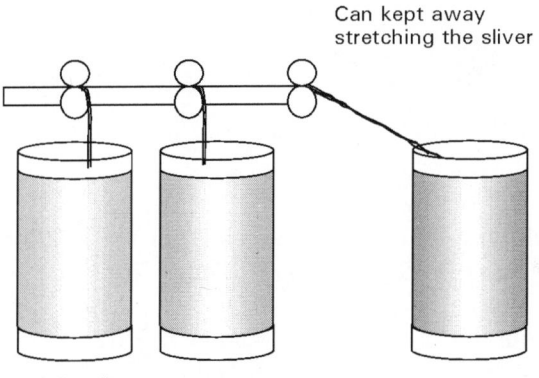

Can kept away
stretching the sliver

4.12 Feeding cans.

3. **Overlapping the slivers while piecing in the creel:** Some of the workers have a practice of overlapping the slivers while creeling as it is easy; but they are unaware of the consequences. At the overlapped place, the sliver thickness shall be high, and it lifts the top roller. The sliver adjacent to the overlapped portion does not get the required pressure, and the drafting shall not be uniform (Figs. 4.13 and 4.14).

4. **Cutting the sliver in cans at creel and reversing for adjusting the cans in creel:** In case of any shortage of material for draw frames, it is a normal practice in number of mills to cut the cans and adjust them so that the draw frames work. But the cutting of cans and lifting the slivers and putting in another can disturbs the layers, and can lead to uneven sliver due to stretching. By reversing the cans, the position of the hooks get changed. In pre combing draw frames, by this action of reversing, we shall be defeating the purpose of having two passages before combing, and shall end with higher fibre loss in noils.

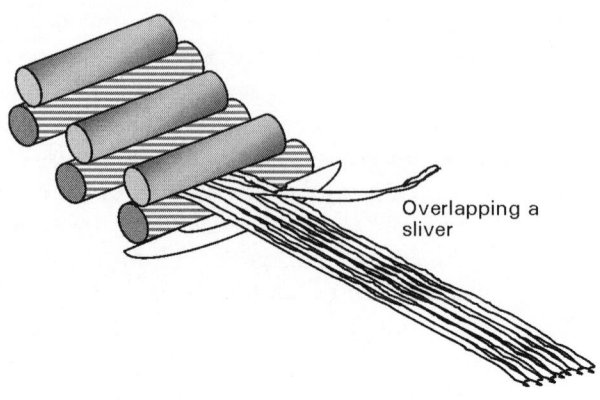

4.13 Creel.

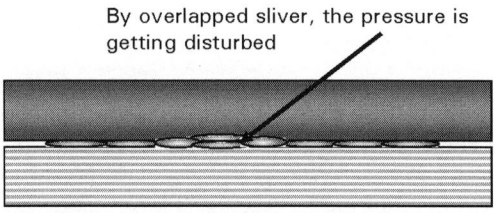

4.14 Overlapping of sliver.

5. **Not removing the sliver wastes from the spring bottom before putting cans in the machine:** Sliver wastes and fluff are normally seen accumulated below the springs in a can. It is needed to remove them before placing an empty can for filling. If it is not cleaned, the dust and fly in the can get contaminated as the air from below the spring moves upwards as the sliver is filled and spring moves down.

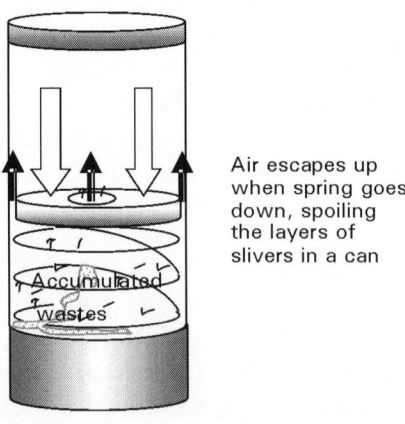

4.15 Accumulated wastes in cans.

6. **Not removing the slivers from the can whenever there is a lapping:** Lapping does not take place in one stroke. First a small portion of cotton sheet gets lapped and then it carries the neighbouring fibres and the sliver breaks. If we analyse the sliver before lapping, we can see 8–9 m of singles, that too in a tapered way. Therefore, whenever there is a top roller lapping, minimum 10 m of sliver should be removed from the can.

7. **Making the stop motions inactive:** Workers inactivate the stop motions while attending to certain repair or maintenance activities, but might not reactivate them might be due to ignorance, negligence or consciously. As there are no leveling points after draw frames, ensuring the working of stop motions is very essential. It shall prevent singles and doubles in sliver, which otherwise would lead to count variations, unevenness and strength variations.

8. **Not getting the cots buffed in time:** Periodic buffing of top roller is very important to get the required quality. The frequency depends on the type of materials being produced. In case of polyesters we need frequent buffing, whereas for combed cottons once in 20 days (maximum 30 days) is practiced.

9. **Replacing the top rollers without verifying the diameter and surface condition:** The diameter and the surface condition of top rollers are very important to get quality slivers. After each lapping, the tenter should check the condition of the top roller. While replacing the top rollers, care should be taken to see that the diameter of all the top rollers are same, and as per requirement.

10. **Not greasing the top roller bushes in time with proper grease:** The quality of top rollers is very important as it can influence the sliver quality directly. The top rollers have two bushes on both the sides. Fine fluff particles can enter the bushes while working and the grease become dry. This can result in jerky motion of the top rollers, and also can damage the top roller surface. It is therefore needed to clean and grease the end bushes of top rollers on a regular basis.

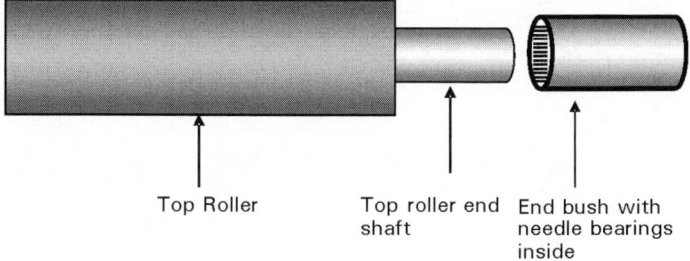

Top Roller   Top roller end   End bush with
                    shaft   needle bearings
                                        inside

4.16 Top roller bushes.

11. **Using sharp knives to remove lapping resulting in damaged cots:**
    Lapping is a normal problem in draw frames, especially when fine cottons
    and long staple fibres are in use. It is difficult to remove the lapping by
    hand especially in the case of staple fibres. The workers invariably use
    sharp knives to remove the lapping. The use of sharp knives can cut the
    rubber surface and make the cot ineffective. It is always suggested to
    keep spare top rollers on each machine, so that the worker will start the
    machine immediately, and remove the lapping leisurely.

Cut mark on cot due to using of
sharp knife to remove lapping

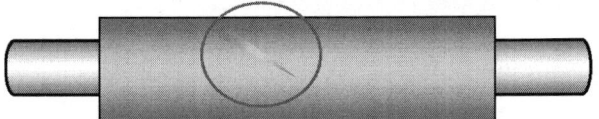

4.17 Removal of lapping.

12. **Not removing the suction box wastes in-time:** Suction boxes are
    provided to suck the liberated floating fibres from drafting zone. The
    wastes are collected in a suction box. The workers need to periodically
    remove the wastes. If the wastes are not removed, it will cover the
    complete mesh, and the suction becomes ineffective. The floating fibres
    fall back on the drafted materials leading to a bad quality.
13. **Keeping the suction box doors half open:** In a number of cases, it
    is seen that the doors of suction box are not closed fully after removing
    the wastes.

Suction duct

Air entering
from door

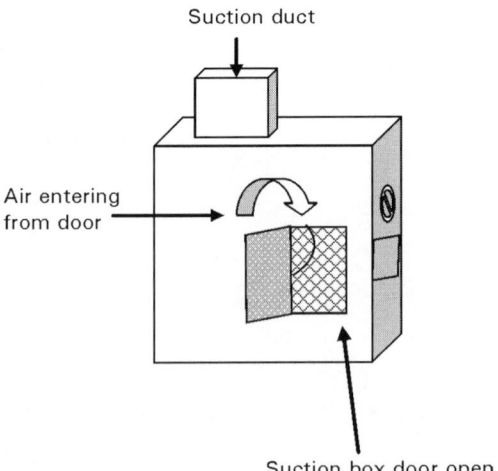

Suction box door open

4.18 Suction box.

14. **Putting cotton pads to get pressure on drafting rollers:** In some cases, due to wearing out of saddles, the pressure on the top roller reduces, leading to uneven drafting and higher breakages. The workers normally put some padding, may be a bit of cotton or card board to get the pressure, and run their machines. But this practice can spoil the machine further. It is always recommended that the fitters attend the machine and replace the worn out parts without undue delay.

15. **Using damaged cans and springs:** The damaged cans and springs disturb the slivers and splitting takes place. This is more dangerous in draw frames, as there is no correction point after draw frame. Therefore, damaged or dented cans and springs should not be used. All damaged cans are to be removed out of the work place, as the workers might use those cans that are available in the work area. The shop floor management should ensure that all broken cans are promptly removed and discarded and the workers are educated suitably.

16. **Using nails and hammer to remove cotton from jammed trumpets:** It is seen in a number of cases, sharp nails and hammer are used to remove jams from a trumpet. It is suggested to use softer materials in the nail, so that the trumpets are not damaged. A damaged trumpet spoils the sliver and increases breakages.

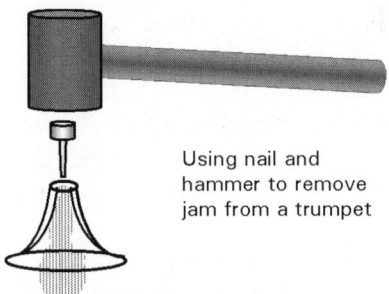

Using nail and hammer to remove jam from a trumpet

4.19 Use of nail and hammer.

17. **Using trumpets without verifying their size and condition:** The size of the trumpet should be such that it compact the sliver, but do not choke up. A wider trumpet gives lesser U% compared to a narrow trumpet, but the strength of sliver reduces. If the extended creels are used in the next machine, and the speed of feed is high, there might be breakages and stretching of slivers. A narrow trumpet although results in a higher U% of sliver, as it does not allow stretching of slivers, gives lesser variations in next process, and hence it is preferred. Care should be taken to see that chocking is not there. If the sliver is uniform, we can comfortably go for a smaller trumpet.

18. **Not ensuring proper alignment of can:** Alignment of can with coiler is very important. If the can is not aligned, the slivers shall fall out of the can and get folded as the can fills up. This shall give problem in taking out the sliver in the next process. The sliver might stretch or break leading to poor yarn quality.

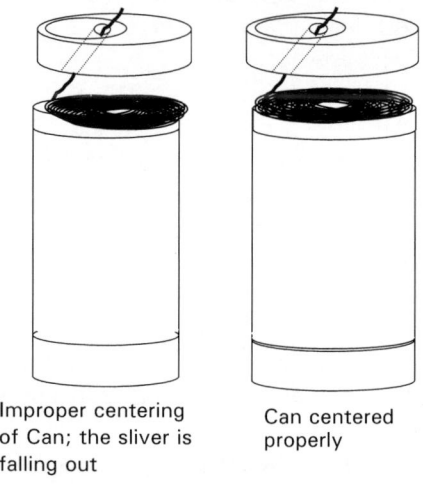

Improper centering
of Can; the sliver is
falling out

Can centered
properly

4.20 Proper alignment of cans.

19. **Not following the correct colour codification and channeling:** Colour codification has a purpose of easy identification of material in process and ensuring that right material is used all the time. Non-following of colour codification for any reason can lead to mix ups and variations in yarn quality. Also tracing back incase of problems shall not be possible if proper colour codification and channelizing is not followed.

20. **Over filling the cans:** Over filling of cans results in sliver splitting in the next process because of the fibres sticking to the adjacent layer.

21. **Pressing the sliver in can by hand:** Pressing the sliver by hand disturbs the coils and can lead to stretch in sliver and count variations.

22. **Use of compressed for cleaning while machine is working:** By using compressed for cleaning while working, we are adding contaminations and also the machine starve for required air pressure, leading to malfunctioning and poor quality.

23. **Un-staggered sliver piecing:** If group creeling is used, care should be taken to see that the entire sliver piecing from the creel should not enter the tongue and groove roller at the same time.

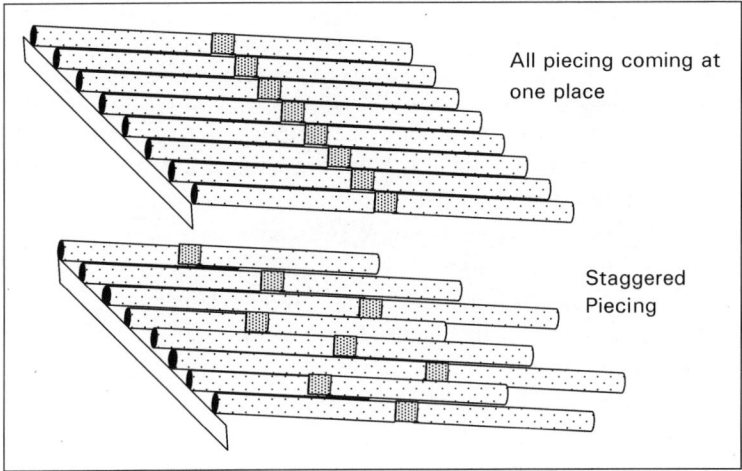

4.21 Unstaggered sliver piecings.

24. **Checking the wrapping periodically and correcting the hank by changing the pinion:** It is a normal practice to check the wrapping of draw frame slivers 3–4 times in a shift and changing the draft change pinion to correct the wrapping. The intention of this exercise is to reduce the count variations and to maintain the average count as per standards. But this system itself is introducing variations. We check 5 or 10 yards of sliver and take a decision. If the hank is heavy we make it light, and if it is light, we make it heavy. But what happened to the material already produced, which is working in the speed frames? Are we going to keep the material separate after every pinion change? No. We are mixing the material produced with different pinions in the same speed frame creel that has materials produced in the last 4–5 shifts. We check wrapping at periodic intervals, but when the wrapping shifted is not known to us, and also how long that shifted wrapping remains is also not known. Therefore, by checking the wrapping and changing the pinion, we are introducing more variations rather than correcting it. We need to put the pinion as per the required draft and ensure that all the machine parts are in perfect order and functioning properly, the humidity and temperature conditions are perfect and the workers are well trained to prevent variations in processes. Thanks to the development of autolevellers that have system of continuous monitoring and correcting of draw frame sliver, because of which this problem of wrapping at periodic intervals are eliminated.

4.2.5 Combers

1. **Bigger laps:** Taking very big laps to get more production in lap forming machine or to avoid stoppages due to spool shortage can result in lap licking problem in combers. This is because of high pressure applied on the inner layers of the lap where fibres have been made parallel by drafting.

2. **Applying excessive pressure to take compact laps:** The laps fed to combers have undergone drafting and the fibres are made highly parallel in order to have lesser hooks. The purpose is to avoid long fibres getting into noils. Therefore, a slight extra pressure on the lap can make the fibres stick to another layer and result in lap licking. Therefore, when the pressure applied is more on the laps, we get lap licking in combers.

3. **Use of worn-out bare spools in lap forming machine:** Lap forming spools if worn out can make the fibres stick to the surface, and also can give jerks while unwinding.

4. **Allowing overlapping of slivers being fed to lap former:** Overlapping of slivers make the lap thicker at that place which leads to loading on half laps.

5. **Overlapping the lap sheet while piecing:** Overlapping of the lap sheet while feeding leads to loading on the half lap and poor combing quality.

6. **Overlapping the slivers on table for piecing:** By overlapping of sliver, the pressure on the adjacent sliver shall be less while drafting giving cuts in the sliver.

7. **Not cleaning the top combs in-time:** The top combs of the earlier models were not having continuous cleaning system as in the latest combers. There is a need to clean the top comb every hour.

8. **Not removing the accumulated comber noils in-time:** The comber noils collected in the box needs to be removed from time to time. It should not be allowed to fill up and press the aspirator drums, as it might damage the surface by pressing.

9. **Keeping tools, etc., on the sliver table:** The sliver table should be always smooth. By keeping tools on the table, scratches might develop, leading to hooked fibres in the sliver.

10. **Inactivating the stop motions:** The stop motions need to be active all the time. If the sliver is uneven, the table trumpet either gets chocked, fall backward or forward depending on the sliver density variations. Hence the machine stops frequently and production shall come down. It is seen that workers make the stop motions inactive in case of more stoppages due to uneven sliver. They need to get the back materials and the machines attended rather than inactivating the stop motions.

11. **Piecing the slivers by over twisting:** While piecing the broken sliver, they need to be gently pressed and slightly twisted. If the pressing and twisting is more than required, we get problem of undrafted material in the next process or thick and thin places.

12. **Not following the correct colour codification and channeling:** Colour codification is very important to ensure that the materials move in correct path and do not get mixed up with other materials. Channeling helps in tracing back the machines whenever there is a problem.

13. **Over filling the cans:** Over filling the cans results in fibre sticking to adjacent layers and split while taking out from the can for next process.

14. **Using damaged sliver trumpets:** Damaged trumpets increase breakages in combers, and also produces hooked fibres in sliver.

15. **Pressing the sliver in the cans by hand:** Pressing the sliver by hand disturbs the coils and can lead to stretch in sliver and hank variations.

16. **Use of compressed air for cleaning while machine is working:** By using compressed for cleaning while working, we are adding contaminations and also the machine starve for required air pressure, leading to malfunctioning and poor quality.

4.2.6   Speed frames

1. **Cutting the slivers at cans in the creel and distributing to other cans:** It is a normal practice in number of mills to cut the cans and adjust the slivers on all working spindles to avoid partial running of machines in case of sliver running out or during shortage of back material. By cutting the slivers and lifting by hand, the layers get disturbed leading to higher breakages in speed frames and ring frames. In such cases, it is suggested to set the draw frame counter to get lower length of slivers.

2. **Overlapping the slivers while attending a sliver run out:** By overlapping the slivers, we get heavy material leading to hard twisted rove, count variation and breakages in speed frames as well as in ring frames.

3. **Cross creeling of cans:** The cans should be arranged in such a way that the slivers are taken in a straight and shortest path. Cross creeling of cans lead to stretch in sliver resulting in count variations at ring frames (see Fig. 4.22).

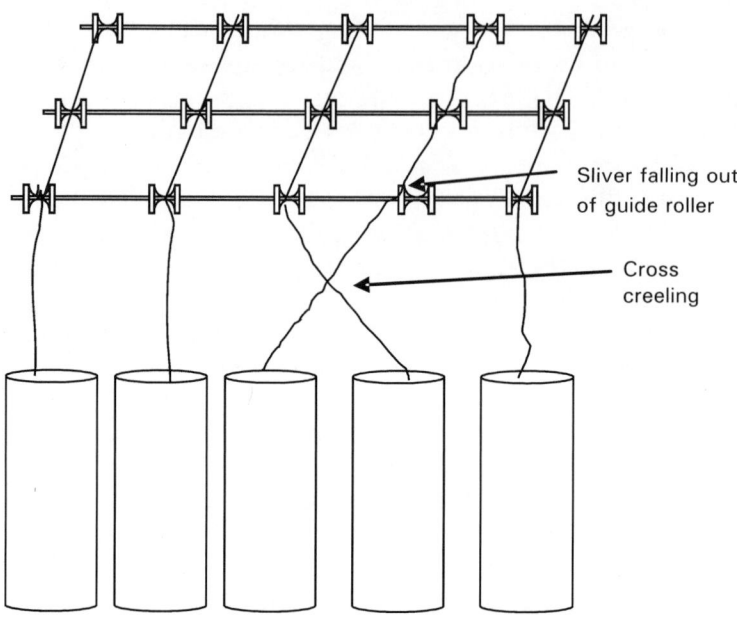

Sliver falling out of guide roller

Cross creeling

4.22 Cross creeling of cans.

4. **Sliver falling out of guide roller:** By not adjusting the sliver on the guide rollers in the creel and allowing them to fall to the side creates problems of stretch. The speed frame creels are normally positively driven, and the tension draft is calculated by taking the diameter of guiding rollers into consideration. If the sliver falls out, then they get higher stretch resulting in count variations.

5. **Twisting too hard while piecing:** While piecing the rove, we need to give slight twist. It is seen that workers in a number of cases give very high twist in order to ensure that rove is strong at the place of piecing. But this results in undrafted ends in spinning leading to breakages or uneven yarns.

6. **Applying cotton to the top arms to get pressure:** Whenever there is low pressure in top arms, workers put some padding of cotton to get the pressure and work the spindles. But this shall damage the top arms permanently as there is no control on the amount of increase in pressure.

7. **Pressing the front top cot to correct a loose end:** We get loose ends in speed frame if a break is not attended in time, as the diameter of the roving bobbin shall become less compared to other good bobbins. In such cases, the workers press the front top roller gently for sometime and see that the tension is made up. This action damages the front cot as the bottom roller will be revolving. Also it creates unevenness in the rove.

8. **Not verifying the spacers and condensers while replacing:** Care should be taken to see that the spacers and condensers are verified before replacing. If wrong spacers and condensers are used, we get uneven rove.

9. **Not putting back the fallen floating condenser while attending a break:** The floating condensers are provided between the apron bridge bar and the front roller nip. In some systems, two condensers are joined with a pin, whereas some others give individual condensers. These are likely to fall down whenever there is a break in sliver. The workers attend the break, but do not put back the condenser in a number of cases, resulting in spreading of drafted rove and creating breakages.

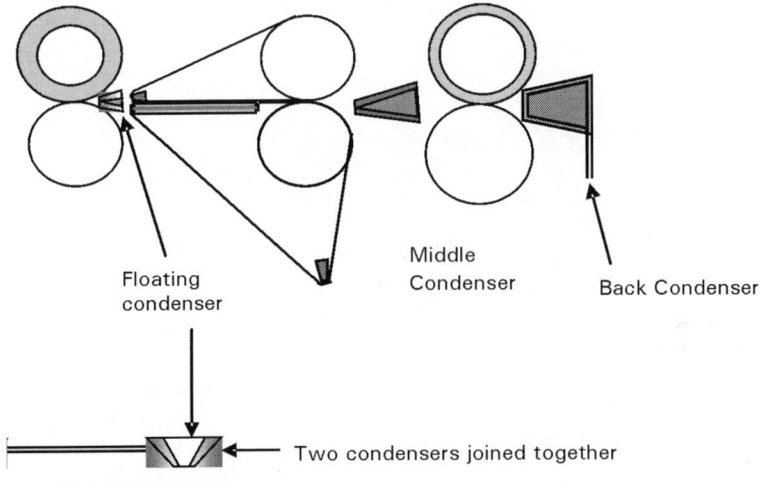

4.23 Floating condensers.

10. **Keeping the top clearer cloth cover lifted above:** The clearers are meant for keeping the cots clean by continuously removing the liberated fibres and wax depositions on cots. While piecing an end, it is needed to lift the clearer cloth or remove the clearer roller depending on the type of breakage. However, care should be taken to put them back inaction. If it is not put back, there shall be problems of lapping on top and bottom rollers (Fig. 4.24).

11. **Removing the separators:** The separators are meant for preventing lashing in whenever an end breaks. However, it is seen that in a number of cases, the separators are not put back into its position after attending a break. As the spindles are working, even the supervisors do not insist on putting back the separators. The problem comes whenever there is a break, as the broken end gets entangled with adjacent end, and spoils material on that bobbin also (Fig. 4.25).

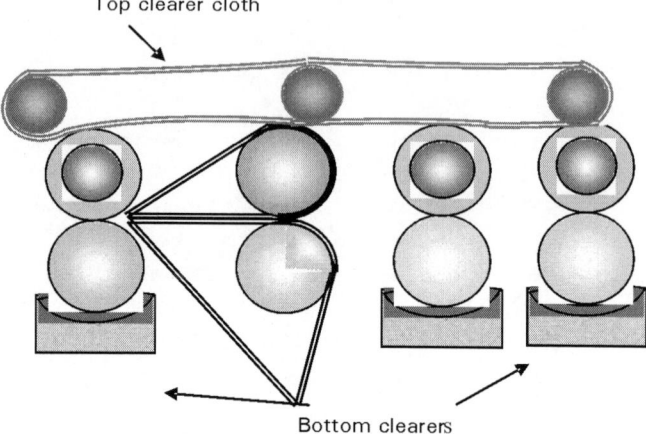

4.24 Clearer cloth.

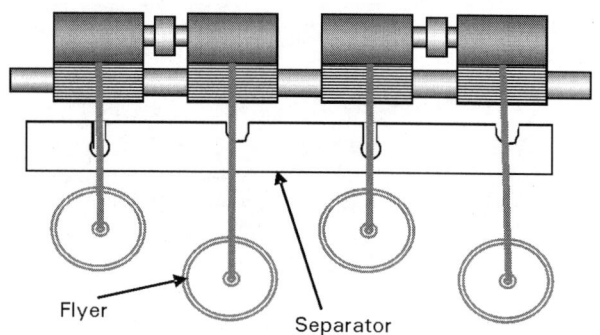

4.25 Separators.

12. **Having uneven wraps on the flyer finger (Presser):** The number of wraps on the flyer presser should be uniform on all spindles. However, in some cases, after a break we get loose ends, and at that time, the worker reduces the number of wrap by one on that spindle. After sometime the tension gets adjusted, but the worker does not correct the warp as he needs to stop the machine to do that. In this process there are chances of him forgetting this. This gives stretch variations, leading to count variations (Fig. 4.26).

13. **Not having the flyers aligned:** All flyers need to be aligned in one position only so that it will be easy for attending breaks and also for doffing. If the flyers are not aligned, at the time of attending a break, we need to inch the machine a number of times, which leads to jerks on all other spindles introducing thin places in the yarn produced form all the spindles (Fig. 4.27).

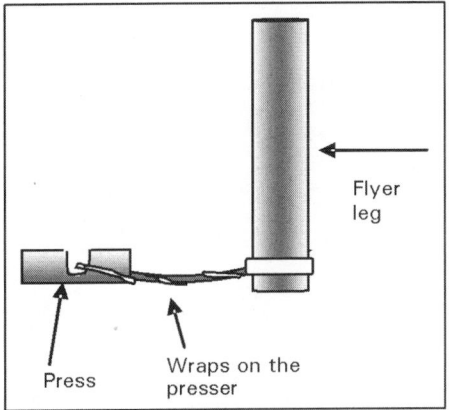

4.26 Uneven wraps on flyer presser.

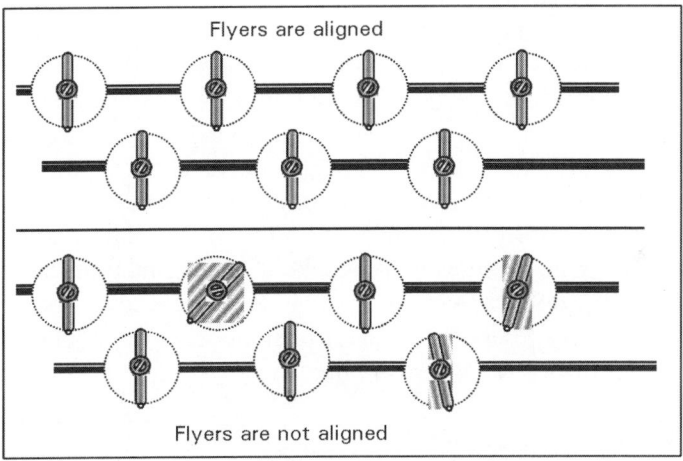

4.27 Aligned and unaligned flyers.

14. **Stopping the machine without verifying the bobbin rail position:** The machine should not be stopped when the bobbin rail is either in the top most position or in the bottom most position. By stopping the machines at these extreme positions, the coils will slip while starting the machine and results in multiple breakages (Fig. 4.28).

15. **Making the stop motions inactive:** In speed frames, the stop motions are normally provided for end breakage in the front, sliver breakage in the back and as a safety devise for the doors. Sometimes, due to fluff accumulations are not cleaned, the stop motions act although the sliver has not broken. In such cases, the workers tend to disconnect the stop motion instead of cleaning the fluff from the sensor area.

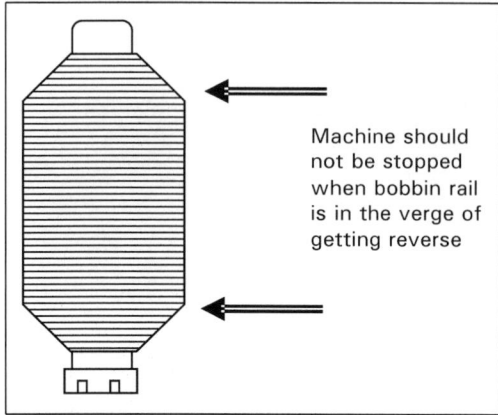

Machine should not be stopped when bobbin rail is in the verge of getting reverse

4.28 Bobbin getting reverse.

16. **Keeping the doffed bobbins on the top arm cover while the machine is working:** At the time of doffing a speed frame, the bobbins are kept on the top of clearer roller cover, i.e., in between the top arms, but the doffers should remove all the bobbins from the machine top before starting the machines. It is seen that when two machines come for doff at the same time, in order to start the machines faster, the doffer boys are not allowed to remove the bobbins from the doffed machine, but are taken for doffing the next machine. The frame tenter starts the machine as he is more worried about his production, and does not insist on the doffers to remove the bobbin before starting the machine. In this process, there are chances of some loose ends interfering with the running ends and creating breakages.

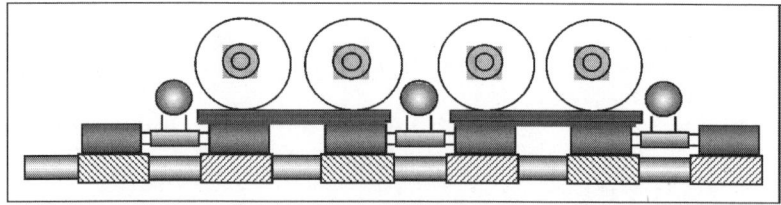

4.29 Doffed bobbins kept on the top clearer plate, not removed after the machine is started.

17. **Not removing the accumulated clearer wastes on time:** The clearer wastes accumulated are to be removed from time-to-time or else they can fall again on the drafting rollers and get contaminated with the rove.

18. **Not following the correct colour codification and channeling:** Colour codification is meant for ensuring correct material getting fed. Hence any deviation can lead to mix ups.

The present day trend is to cut off the sliver and remove the bobbin from speed frame in case of a break instead of earlier practice of piecing the rove. The concept is to avoid sliver piecing or roving piecing from speed frame working in ring frame as they result in thin and thick yarn not only from the bobbin that had a break, but also from all the other bobbins running on the speed frame at that time. One might argue that the defects get cut the yarn clearer in winding, but, we should know that all yarn faults created by piecing are not cut. It might go to fabric. We need to work towards zero breaks in speed frames to avoid problems of yarn unevenness and weak spots.

4.2.7   Ring frames

1. **Keeping roving bobbins on the top of creel for long time:** There is a provision to keep roving bobbins on the top of the ring frame creel to facilitate the tenter to replace the bobbins as and when they run out. It does not mean that the creel top can be used as a stacking place for bobbins. By keeping bobbins for a long time on ring frame creel, the dust accumulation shall be there, and also the bobbins become soft. The bobbins are to be kept is separate racks or trolleys.

2. **Overlapping of roves while replacing:** When a bobbin is running out, the workers normally cut the old bobbin and creel the new one, and insert the rove into the traverse guide while still the old rove is working. This is done to avoid the breakage. But care should be taken to see that the overlapping done is removed immediately after the fresh rove passes through the traverse guide and enters the nip of back rollers. If it is not done, we get long thick yarn of double the density. It is normally seen that the workers allow overlapping of 15–20 cm, which gets drafted by 25–35 times and we get a long thick fault (see Fig. 4.30).

3. **Putting cotton in top arms for getting more pressure:** Due to various reasons like disturbed setting of saddles in a top arm, improper alignment of springs, improper setting of top roller, air leakages in case of pneumatic loading, etc., we get lower pressure in the top arm. The workers, in order to ensure that the spindle works, put a pad of cotton and get the required pressure for drafting. However, this practice can result in permanent deformation of top arms, as all workers cannot understand and manipulate to get the required pressure. In a number of cases it shall be excess pressure. It is always suggested to get the help of fitter and ensure that the top arms are set properly.

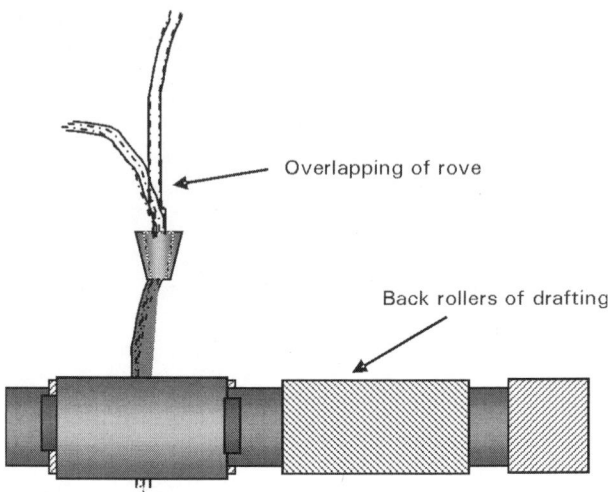

4.30 Overlapping of roves.

4. **Using sharp knives/hooks for removing lapping:** It is found that workers use sharp objects like pieces of licker-in wire, knives, etc., to remove lapping on top rollers. By this the cots get damaged. The workers need to be given some extra set of top rollers so that they can replace the top roller and start the spindle, and clean the top roller later by hand.

5. **In active bottom clearers resulting in partial lapping:** It is normally seen that the bottom clearers are not maintained in number of mills, especially for back bottom rollers. Although the back bottom roller runs at a very slow speed and there are less chances of cotton lapping compared to front rollers, we can see wax deposition on them, and sometimes a fine layer of cotton fibres lapping partially. As it does not create a break, it goes unnoticed. However, we need to understand that the lapped portion is bound to lift that portion of cot, resulting in reduced pressure acting on the material being drafted. Hence it is bound to result in a lower draft and also uneven draft (Fig. 4.31).

6. **Not attending a break for a long time:** The breaks are to be attended immediately or else it might result in multiple breaks due lashing in and fibres liberating from the running cop (Figs. 4.32 and 4.33).

7. **Cross-creeling:** Cross-creeling in ring frames can lead to stretch. Hence there shall be higher variations in count (Fig. 4.34).

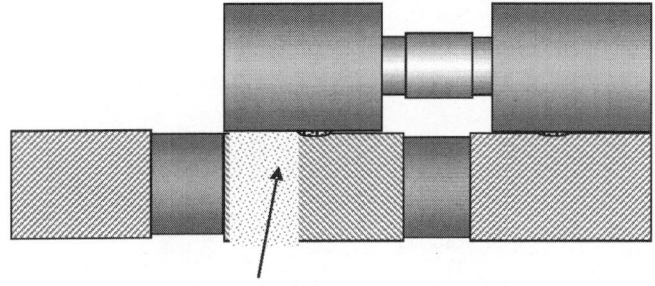

Partial lapping on back bottom roller
The top roller is lifted slightly

4.31 Partial lapping.

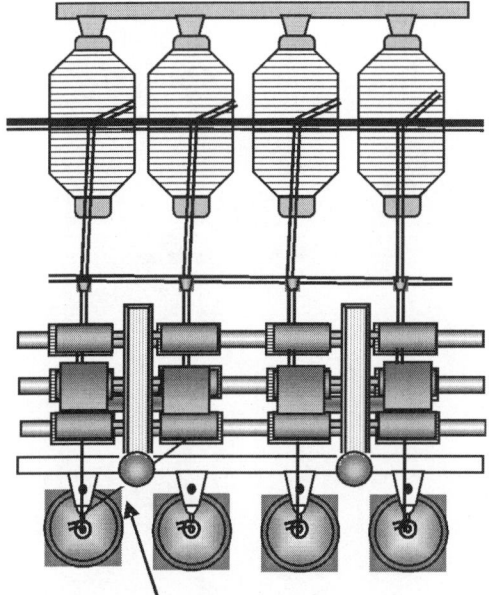

Broken end lashing into
adjacent spindle

4.32 Broken ends.

8. **Not removing the last layers in the roving bobbin and allowing it run out fully:** In conventional speed frames the flyers are fitted on the spindles. In such case, before stopping the machine for doffing, the drive to bottom cone drum is disconnected, and the drafted rove is allowed to get accumulated on the top of the flyer. While doffing, the flyers are lifted up and the bobbins are taken out and the new bobbins are put. In this process, the loose rove, which has not sufficient twist, gets stretched. Afterwards, the rove is wrapped on the fresh bobbin, and the machine is started. In this process, the

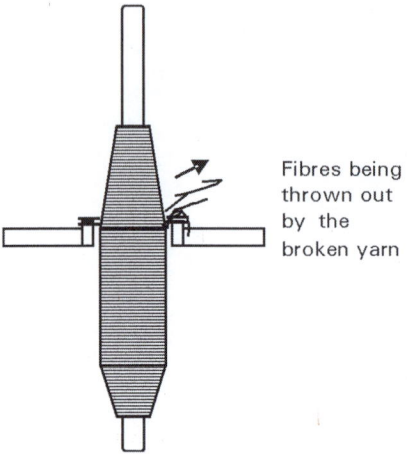

Fibres being
thrown out
by the
broken yarn

4.33 Broken yarn.

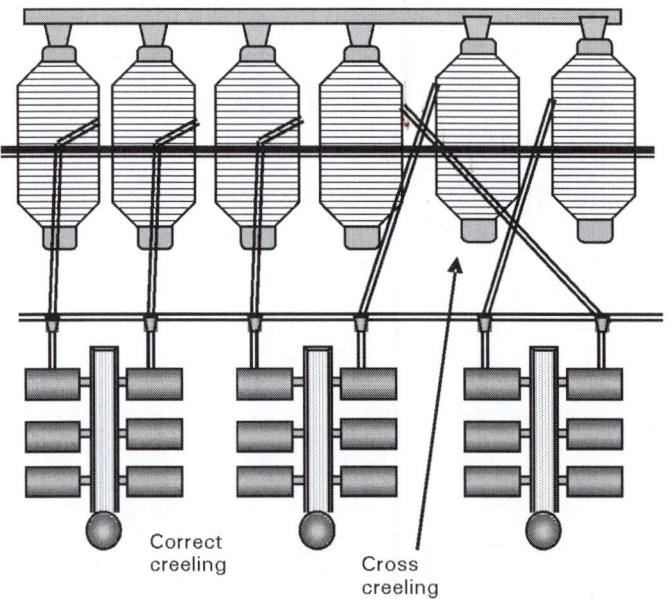

Correct
creeling

Cross
creeling

4.34 Correct creeling.

material in between the front roller nip and the bobbin, approximately
1.5–3 m are very irregular. Therefore, it is better to remove the last
layers from the roving bobbins on the ring frame creel rather than
running them to reduce the roving wastes (Fig. 4.35).

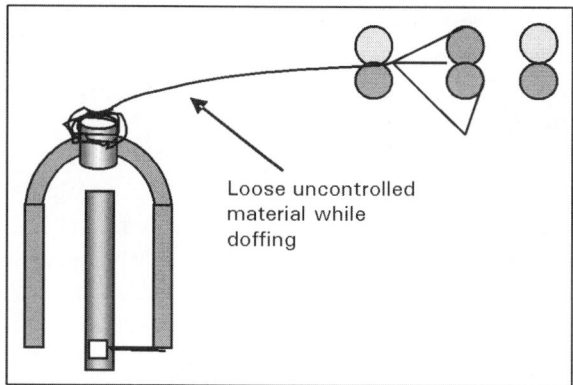

4.35 Loose uncontrolled material.

9. **Keeping roving bobbins in the path of roving or on the top arms:** In some mills, it is seen that the workers keep the bobbins to be creeled in the path of roving, instead of keeping it on the top of the creel. The reason is that the height of ring frame creel is more. The problem is more with the machines designed in European countries, where the height of normal people is more than the height of Indians (Fig. 4.36).

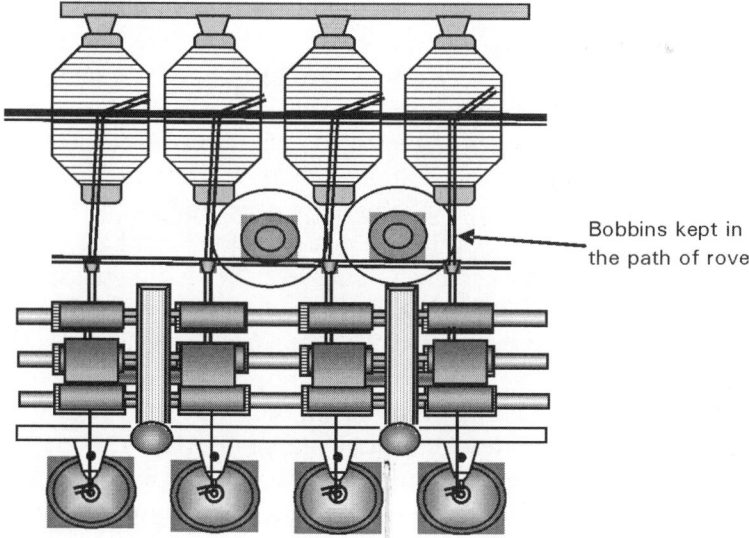

4.36 Roving bobbins.

10. **Keeping the bobbins in the path of air flow of overhead cleaners:** The overhead cleaners are meant for preventing the loose fly from not settling. One nozzle blows the air and the other sucks the loose fibres.

The positioning of the nozzles is very important. Bobbins should not be kept in the path of the air flow, as the air flow can make the fibres in the bobbins loose. Hence it leads to thin places in the yarn.

11. **Not removing the bobbins while a tape is cut:** When a tape is cut, the spindles are not stopped immediately. It might take around one minute for the spindles to stop. The speed of the spindle shall be gradually coming down, and the end shall break when the twist is very low and not sufficient to keep the yarn in tact. Therefore, the twist shall be very low in all the four cops those were being run with that broken tape. It is suggested to remove the cops from the ring frame spindles, and clean 15–20 m of yarn and then put back for piecing. Otherwise we might get a problem of dark line, especially in light dyed fabrics.

12. **Not replacing the travellers in time:** The travellers shall be running continuously on the flange of the ring. They get worn out depending on various factors like the speed, the count being worked, the material being spun, etc. The traveller needs to be changed before it burns. The studies of hairiness index and imperfects show that after 2 or 3 days working, the imperfections start gradually increasing along with the hairiness.

13. **Not following the running-in schedule of rings properly:** The running-in procedure is decisive for the future positive/negative behaviour of the ring and the length of its service life. Every ring requires a certain degree of running-in time if it is to maintain high traveller speeds with as little ring and traveller wear as possible. Geocities.com[5] suggests that during running-in the use of steel travellers without surface treatment is recommended. After the termination of the running-in process, steel travellers with surface treatment or nylon as well as bronze travellers can be used.

　　The running-in process, beginning with the starting phase, consists of improving the initial running properties of the metallic running surface up to the optimal values by smoothing and passivation (oxidation) as soon as possible. In this way, together with fibre lubrication, constant minimum mixed friction conditions and minimum thermal stressing can be attained for the ring traveller. A careful running-in process will improve the lifetime of the rings. In order to keep the stress on the traveller as low as possible during the starting phase, it is advisable to always change the traveller in the upper third part of the cops. Further advantages are brought with the use of a traveller running-in programme (reduction of the speed by about 10% for 10–20 min, only available on modern spinning machines).

Spindle speed should be reduced at least for the first 10 traveller changes. If final speed is higher than 32 m/s, reduce by at least 20%. If final speed is lower than 32 m/s, reduce by at least 10%. New rings should not be degreased, but only rubbed over with a dry cloth.

The running in should be done with the same traveller type which is used for normal operation with the 10–20% less than normal speed. It is not advisable to do running with the same speed but with 1–2 numbers lighter travellers than usual. The first traveller change should be carried out after 15 min. The second traveller change should take place after 30 min. The third traveller change should be made after 1–1.5 h. The fourth traveller change should be made after the first doff. Further traveller changes are to be made according to the manufacturers' recommendations.

14. **Keeping excess travellers in the cup and not removing them while changing counts:** Normally cups are provided in a ring frame to put some extra travellers, so that in case of a traveller fly, the same can be attended. However, it is seen in a number of cases, that when counts are changes, all the old travellers are not removed from the cup. Hence the worker uses the traveller in cup, leading to uneven tension. This might lead to higher breaks and increase in hairiness (Fig. 4.37).

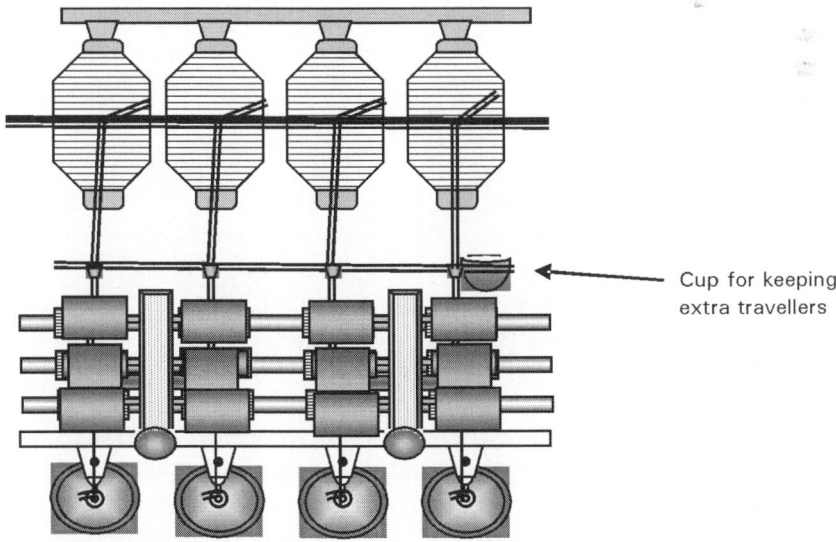

Cup for keeping extra travellers

4.37 Excess travellers.

15. **Not cleaning the rings with dry cloth/cotton before replacing a burnt traveller:** Whenever a traveller burns, we see deposition in the inside portion of the ring. It should be cleaned with a dry cloth before replacing the traveller, or else, the new traveller also burns very fast. Further the hairiness shall be high.

16. **Disturbing the traveller clearer settings:** The traveller clearers have a function of keeping the travellers clean. If it is not set properly, the traveller gets loaded with fluff and the weight of the traveller increases. This results in higher breakages in ring frames, and also higher hairiness in the yarn. It is normally seen that the people do not give attention to this setting. The setting normally gets disturbed while centering the rings. We need to take the traveller clearers back for adjusting the rings, but should remember to set it back after the rings are set. The other reason for disturbed setting is rough handling, especially while doffing manually. The bobbins sometimes slip and fall on this and the clearer bends resulting in disturbed setting (Fig. 4.38).

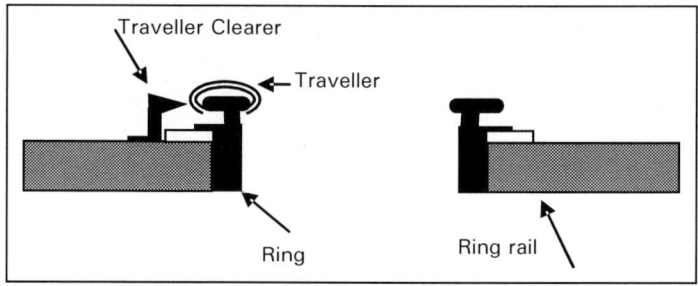

4.38 Traveller clearer.

17. **Holding spindle for a long time while piecing ends:** Holding the spindle for a long time results in resistance to the movement of tape, and reduces the speed on other spindles being driven by the same spindle tape. Thus variations in twist are introduced in other spindles.

18. **Holding yarn for a long time while piecing:** By holding the yarn for more time than required, we end up with very high twist in that portion of the yarn.

19. **Putting a long tail while piecing:** A long tail end while piecing leads to a bad piecing. These defects are classed under C and D faults as per Classimat. The tenter is normally held responsible for bad piecing, but it is equally important that the end breakages are maintained at a minimum level so that the tenter can do good piecing comfortably (Fig. 4.39).

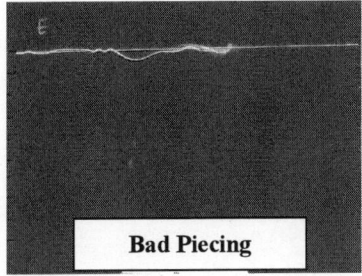

4.39 Bad piecing.

20. **Piecing from over the cot:** While piecing a broken end the worker need to piece at the nip of the front rollers. However, it is difficult in case of very high front roller speeds and with less trained workers. They tend to take the end over the front top roller and insert it from the back at the nip of aprons feeding the materials to front rollers. Because of this we get a long bad piecing, that is bound to get cut in the winding clearers reducing the winding efficiency and also higher hard wastes. The workers need to be properly trained before putting them on the ring frames for regular production (Fig. 4.40).

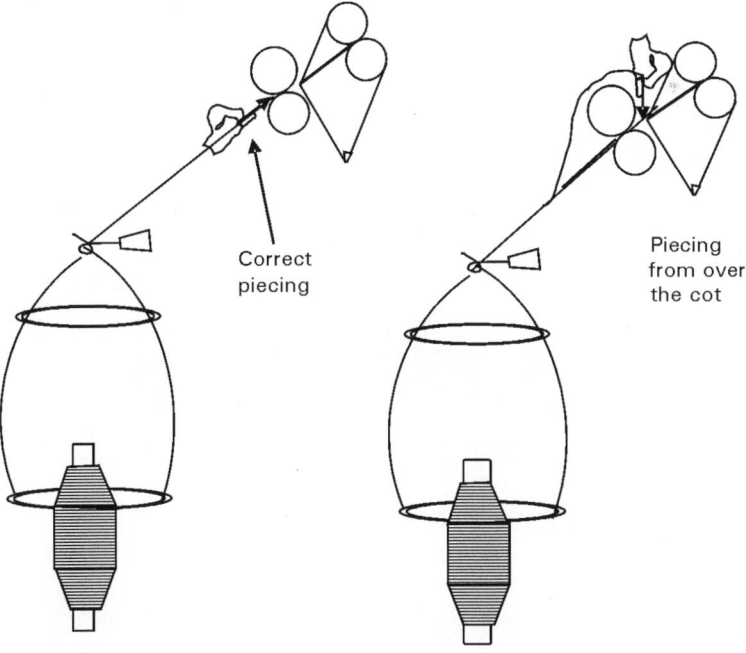

4.40 Correct piecing.

21. **Holding the bobbin by hand for stopping a spindle:** Some workers have habit of holding the bobbins instead of holding the spindle while piecing. By this the coils gets disturbed and might give breakages in winding. Also, it might cause injury to the worker. It is always suggested to use the knee breaks rather than using the hands for stopping the spindles. When the speeds were less than 10,000 rpm, it was easy to stop the spindles by hand, but now, the speeds going above 17,000 rpm and reaching up to 25,000, use of hands should be completely stopped.

22. **Double piecing at cop bottom:** When the end breakages are more in the bottom position in a ring frame, the workers tend to piece the ends by introducing extra end from a spare cop rather than stopping and searching for the yarn from the running cop. It is difficult to hold the spindle as the ring rail will be obstructing the hand movement. The speeds and the settings should be adjusted in such a way that breakages are less in the bottom position, especially after starting the machine after doff. This is also called as double gaiting (Fig. 4.41).

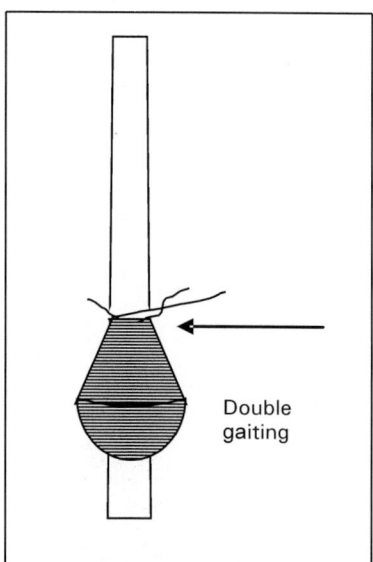

Double gaiting

4.41 Double piecing.

Double gaiting invariably results in discontinuity at winding end. It is therefore a loss of production and also increases wastes.

23. **Taking ring rail down while working to adjust doffs timings:** Sometimes doff comes when the workers are not on spot to take doff, or more number of machines come for doff at a time. In such cases,

where manual doffing system is employed, the jobbers used to take the ring rail slightly down so that the machine can work for some more time. This creates more problems; all the yarn in the top of the cop shall be a waste as it cannot properly unwind. In olden days, in manual winding running at a slow speed of 450 YPM, there was no problem with this practice. Thanks to the new machines, with auto doffing. This problem is completely eliminated.

24. **Not covering the doffed yarn crates in stock:** In ring frames, after the doffing, the cops are stored in crates. These crates should move to winding area, and kept covered till indent comes for yarn from winding; otherwise, fluff shall fall on the cops, which gets caught at electronic yarn clearers creating breakages at winding.

25. **Storing the doffed cops in bags:** Some mills have a practice of storing the doffed cops in bags instead of crates. By this the cops get disturbed and damaged while handling. Also, there shall be a load on the cops at bottom, resulting in soft cops. This shall lead to higher breakages.

26. **Not removing the suction box wastes periodically where rotary filter is not provided:** The sider should periodically remove the wastes collected in the pneumafil box, or else, the suction pressure reduces because of waste accumulation. This shall lead to multiple breaks in case of an end break.

27. **Not cleaning the clearer rollers in-time:** The clearer wastes from the clearer rollers are to be removed periodically, or else the wastes might enter the drafting zone. The clearer rollers might fall down as the clearer roller pins come out of the guides due to excess lapping on the rollers. The clearer roller might fall on the running ends and create more breakages.

28. **Stopping the machine without verifying the ring rail position:** The ring frame should be stopped only when the ring rail is moving downwards or else it shall result in multiple breaks (Fig. 4.42).

29. **Cleaning machine with compressed air while machines are working:** It is seen that workers clean the machines and floors, and also clean themselves by using compressed air. As the machines are working, the fluff gets contaminated with the yarn.

4.2.8   Winding

1. **Bypassing the yarn clearers:** The winders, while knotting a yarn should ensure the passing of the yarn through the yarn clearer after knotting. However, in a number of cases, the worker just put the thread through the yarn clearer and does not ensure it remaining there. The yarn comes out because of improper tension applied or some

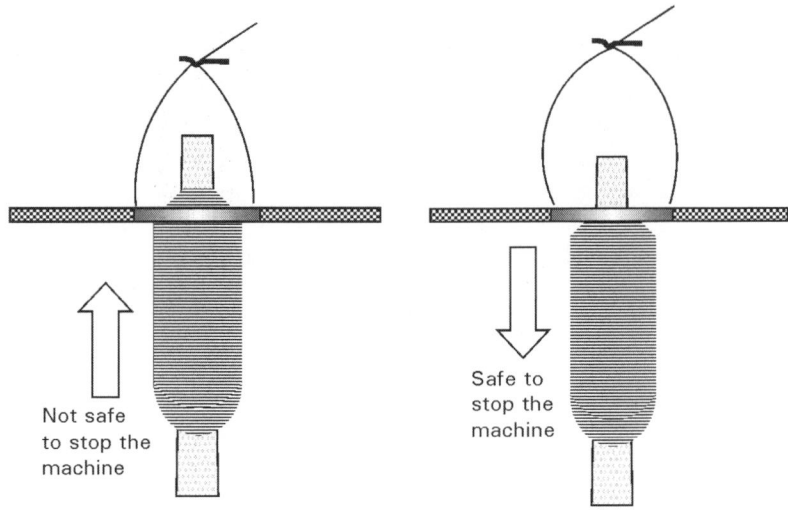

4.42 Bypassing yarn clearers.

obstruction to the yarn. This can happen even in automatic winders. Because of this the winding machine fails to do its basic function of clearing the yarn faults.

2. **Not keeping the clearer area clean:** The EYC area should be always kept clean. In a number of times it is seen that some broken thread shall be hanging near the EYC and sometimes enter the EYC slot. This creates unnecessary breakages.

3. **Keeping tighter setting to get good yarn:** A tighter setting is kept with a good intention. However, it shall create more breakages, and each break requires to be spliced. Sometimes the defect cut might be smaller than the defect produced by a splice. We should therefore verify the yarn quality and the type of faults present, and set the machine to cut objectionable faults only.

4. **Keeping excess tension to get a compact cone:** By keeping extra tension one can get compact cones, but this action reduces the elongation of the yarns. We need to ensure that the elongation properties are not affected while attempting for getting compact cone.

4.3   Machinery condition

The poor quality of yarn from machinery can be attributed to the following:

(a) Worn out parts
(b) Improper maintenance and settings

Let us discuss the effects of poor machine conditions and improper maintenance on the quality of the yarn produced.

4.3.1   Blow room

1. **Burrs in the path of material:** If burrs are present in the path of materials, they shall hinder the free movement of cotton developing hooks and bunches resulting in neps. The inside portion of all ducts should always be smooth.

2. **Blunt grid bars:** Blunt or damaged grid bars hold the fibre tufts and create neps. They also hinder the smooth fall out of trash reducing the cleaning efficiency (Fig. 4.43).

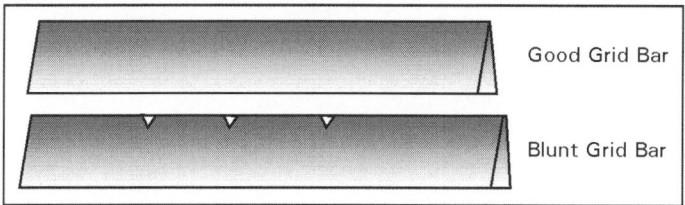

4.43 Blunt grid bars.

3. **Broken or bent pins:** The pins at fine openers like Kirschner beater play a very important role in opening the cottons. If the pins are bent, the cottons get hooked and leads to generation of neps. It is better to remove a pin if found bent or blunt rather than running it (Fig. 4.44).

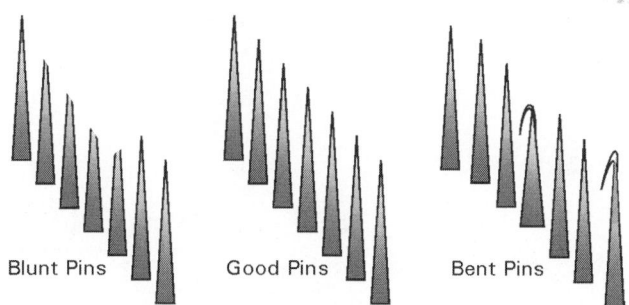

4.44 Broken or bent pins.

4. **Air leakages leading the faulty functioning of valves:** The feed controls and the level controls in the reserve boxes at various places are activated by pneumatic valves. Air leakages result in lower air pressure and hence malfunctioning of these valves. This leads to poor performance of the machine. Materials fed may be more than required or less than required. This may lead to jamming, weight variations, over beating resulting in cat-tails and neps.

5. **Oily belts resulting in slippages:** Care should be taken to see that oils do not fall on the belts, as an oily belt slips on the pulley. The

slippages might result in jamming of beaters, uncontrolled air currents, etc., resulting in a poor performance.

   6.   Too many bends in pipelines: Each bend in the cotton conveying pipeline is a potential area for neps generation. Also the bends restrict the speed of movement of materials demanding for fans of higher capacity and higher power consumption.

   7.   Broken or cracked wooden parts: Broken or cracked wooden parts can lead to contaminations, accidents and break downs of the machines.

   8.   Loose belts: Loose belts result in slippage that leads to frequent jamming, low cleaning efficiency, heat generation on pulleys and possible cause for fire accidents.

   9.   Worn out leather flaps: In blow room line, leather flaps are provided at number of places like strippers in bale breakers and hopper feeders, in delivery boxes, after condenser cages, etc. The basic purpose of these flaps is to prevent cotton from going back. A worn out flap does not take out the cotton out fully and a portion is allowed to go back, resulting in over beating and frequent jamming of the machines (Fig. 4.45).

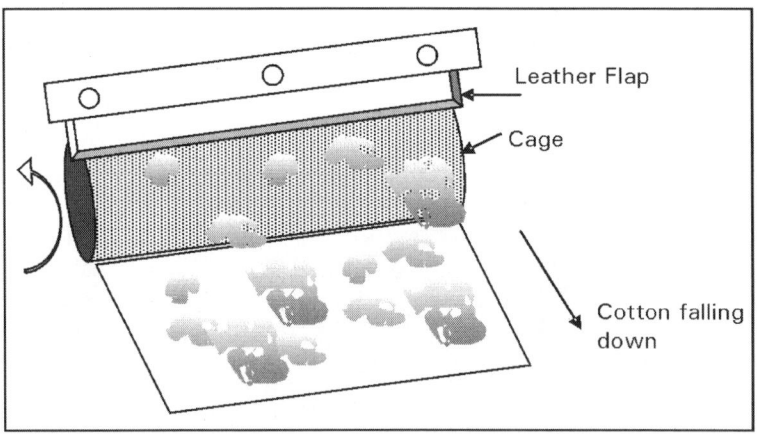

4.45 Worn out leather flaps.

4.3.2   Carding

   1.   Bent or blunt wire points: The condition of the wire points plays a major role in carding. The wires should always be sharp without any hooks or blunt points. Periodic checking and grinding is suggested, which depend on the rate of production and the type of wire points. Vijayakumar[6] suggests grinding frequency for flat tops as once in 3 months for better yarn quality as flat tops plays a major role in reducing neps and kitties in the yarn. He further suggests use of emery fillet rollers for flat tops grinding, instead of using grinding roller grinding stone. Changing of licker-in wires for every 150,000 kg productions in carding, and changing of stationary flats for every 150,000 kg productions in carding is suggested (Fig. 4.46).

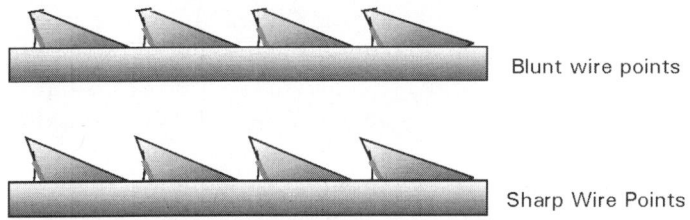

4.46 Blunt and sharp wire points.

2. Worn out flexible bend due to improper lubrication or dirt application:

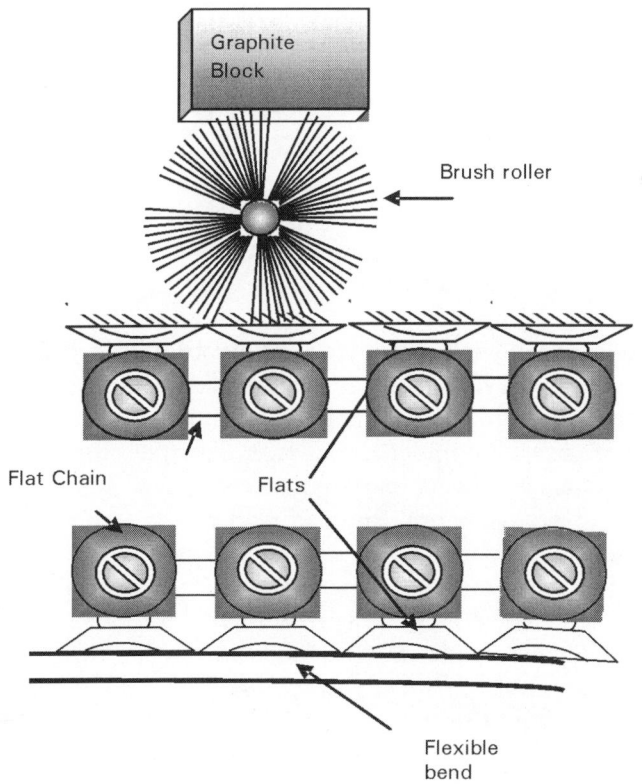

4.47 Worn out flexible bend.

The flexible bend plays a very important role in carding by maintaining the setting between cylinder and flats. Flexible bend is stationary on which the flats move. There is a metal to metal contact. There are always chances of flexible bend wearing out if not lubricated properly. Normally graphite lubrication is provided. However, in a number of cases it is found neglected, and the cards shall be running without the graphite blocks or the brushes not set properly. The brush should be set in such a way that it touches both the graphite block and the heel and toe portion of the flat. The brush should clean the flats, and also apply graphite to the milled portion (Fig. 4.47).

3. **Vibrations due to worn out shafts, bearings and loose fittings:** The worn out parts and loose fittings change the settings in cards leading to a poor quality and also might result in damages to wire points.

4. **Damaged trumpets:** Whenever doubles are produced, chocking takes place at trumpet. While removing the choking, care should be taken to see that the trumpets are not damaged. A damaged trumpet creates hooks and entanglements of fibres in the sliver, which ultimately results as slubs or bunches.

5. **Damaged wheels:** Any damaged wheel gives intermittent motion or jerks leading to poor quality and also to the damage of other parts of the machine.

6. **Loose belts:** Loose belts leads to slips and speeds vary affecting the transfer ratio and carding action. This might also result in loading of cylinders which damage not only the quality of sliver, but might damage the wire points.

7. **Damaged under casing:** Under casings are provided for licker-in and cylinder to prevent good cottons from falling down in the wastes, and at the same time allow the impurities to fall off. It is seen in a number of cases, the under casings get damaged; the rods get bent and the distance between bars increase or decrease, the perforated sheet might get worn out, the side solder might come out, etc. In such cases, the cotton gets accumulated at the damaged portion and obstructs the movement of other cottons, finally leading to a poor quality and damages to costly wire points. The under casings should be checked periodically, at least once in a week for licker-in and once in a month for cylinder (Fig. 4.48).

8. **Loose flat chain:** The flat chain, because of the weight of flats it is bearing and continuous working, becomes loose after a long service. Normally a flat chain works for 3–5 years depending on the conditions. People try to reduce two links in the chain and make it tight. This can only make the chain tight. When the chain becomes

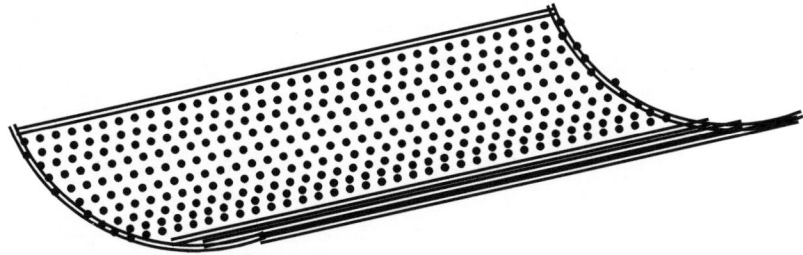

4.48 Under casing.

loose, the gap between two flats shall increase, and the cotton and dusts starts leaking. This might accumulate at any place and spoil the quality. Therefore, when a chain is found loose, it should be replaced, and not worked with reducing two or four links (Fig. 4.49).

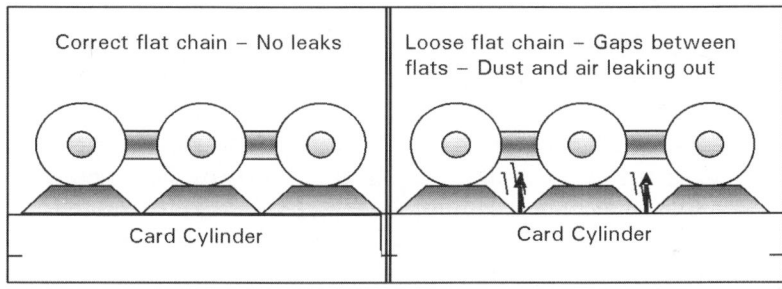

4.49 Card cylinder.

4.3.3 Drawing

1. **Worn out bearings:** Worn out bearings lead to eccentric movement of rollers resulting in short-term variations in drawing sliver. It ultimately ends up as count variation in yarn stage because of the drafts in speed frames and ring frames, which enlarge the defect length by 300–500 times.
2. **Uneven pressures on the top rollers:** Uneven pressures contribute for improper drafting leading to hank variations and ultimately count variation.
3. **Improper buffing of top rollers:** The surface of top rollers should be plain. Top rollers should be checked by the operators at least once in a shift. Also whenever there is lapping, the operator should check the condition of top roller. Periodic buffing is required depending on the type of materials and the speed of working. If the top roller eccentricity is more than 0.05 mm, it should be buffed and top roller eccentricity should be zero after buffing.

4. **Diameter variation between top rollers:** The variation in diameters between top rollers might cause unequal pressures and hence improper drafting. The diameter variation between top rollers should be less than 0.1 mm.

5. **Ineffective stop motions:** The purpose of draw frame is to make the fibres parallel and to make the sliver even. After the draw frame, there are no processes that can level the materials and make them uniform. Therefore, it is very necessary to ensure that bad quality materials are not allowed to move forward. The stop motions provided in draw frames need to be maintained all the time, and the draw frame should not be allowed to work if stop motions are not acting properly.

6. **Air leakages resulting in pressure drops:** Air leakages in the pressure pipes lead to low pressure on the top rollers. This leads to cuts and undrafted materials. This finally leads to count variation in the yarn.

7. **Damaged trumpets:** Damaged trumpets hinder the movement of slivers and results in breaks or hook formation.

8. **Damaged Wheels:** Damaged wheels lead to short-term variations in the sliver. This will lead to count variation in ring frame.

9. **Worn out clearer strips:** Clearer strips are meant for keeping the top rollers clean and prevent from cotton lapping over the cots. Because of continuous working, the rubber in the strips wears out. Also because of heat generated while running, the rubber becomes hard, and does not function as required. The condition of the clearer strips should be checked during every cleaning (Fig. 4.50).

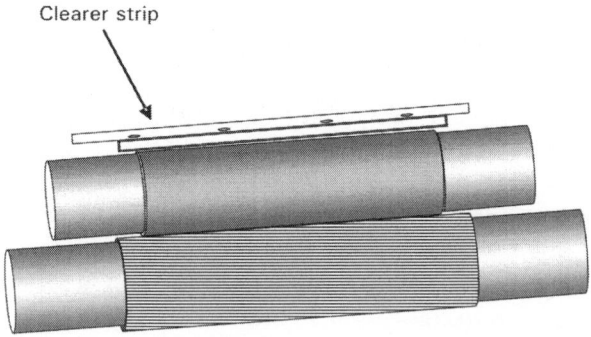

Clearer strip

4.50 Clearer strips.

4.3.4   Combers

1. **Worn out or blunt combs:** Worn out or blunt pints in comber half lap leads to plucking of fibres from nipper grip, and not allowing the fibres to be taken by detaching roller. Ultimately the good fibres are brushed out and taken as noils. If by chance, the fibres are taken by

the detaching rollers, we get hooks and neps, beating the basic purpose of combing. If needles are found damaged on half laps, the entire needle segment is to be replaced. Similarly, the top comb is to be re-needled if found damaged.

2. **Worn out nipper plates:** Worn out nipper plates do not grip the fibres and allow them to move out in bunches. The combing shall not take place and the fibres go in to noil. Nipper grip can be checked by introducing three paper strips of less than 0.1 mm thick across the width of the nipper. It should be corrected by aligning the top nipper blades with the cushion plates and eliminating the worn out parts. Care should be taken that nippers should not work without cotton or else the plates shall be damaged.

3. **Damaged bristles of the brush:** Damaged bristles results in patchy web as the fibres are not cleaned and left on the needles of half-lap itself. Therefore, periodic dressing of bristles is needed. After each dressing, the brush should be set.

4. **Scratches on the sliver table:** The sliver table should be always very smooth as the combed slivers with fibres parallel to the axis of the sliver, move on them. Any rough surface shall lead to hooking of fibres and unevenness of sliver. Therefore, no hard materials should be put/kept on the sliver table.

5. **Improper buffing of detaching rollers:** The detaching rollers in combers have comparatively smaller diameter and a longer length when compared with other top rollers like that of draw frames, comber draw box, sliver lap or ribbon lap. Because of this, when pressure is applied on both the sides of the roller, there are chances of them bowing out. Therefore, taper buffing is recommended depending on the extent of bowing observed (Fig. 4.51).

6. **Damaged trumpets:** Damaged trumpets contribute to hooked fibres, higher sliver breaks, frequent stoppages of comber and higher unevenness of sliver.

7. **Feed roller grip:** Inefficient and uneven feed roller grip results in slippage of fibres during detachment, producing uncombed patches in the comber web and erratic noil levels. The grip should be checked by inserting thin paper strips at different places of the feed roller.

8. **Worn out parts in detaching assembly:** Mechanical defects such as worn out parts in top detaching roller loading assembly, worn out bottom roller flutes, worn out roller bearings, etc., leads to ineffective and uneven grip, slippage of fibres during detaching finally leading to loss of good fibres in the noil. Ineffective grip of front detaching roller results in folding and crumpling of web. Play in bottom detaching roller drive results in alternating stretch and slackness in the web, prominent piecing waves and loss of long fibres in the top comb wastes.

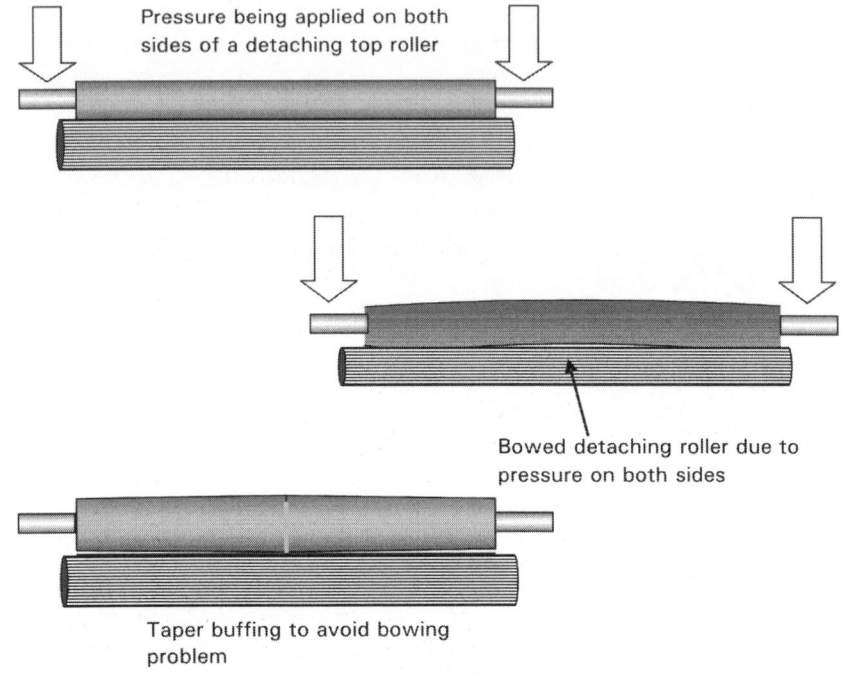

Pressure being applied on both
sides of a detaching top roller

Bowed detaching roller due to
pressure on both sides

Taper buffing to avoid bowing
problem

4.51 Improper buffing in detaching rollers.

9. **Damaged air seals in aspirator box:** Damaged air seals create air turbulence resulting in disturbance of fibres during piecing leading to uneven and patchy web and erratic noil levels.

4.3.5   Speed frames

1. **Damaged wheels:** Damaged wheels in drafting system give cuts and short-term irregularities in roving, which shall become long-term variations in yarn. It also results in higher breakages at spinning, if not at roving. Damaged wheels in twist gears results in higher breakages at roving machine.
2. **Vibrations in flyers and spindles:** Vibrating spindles and flyers give irregular TPI and stretch in rove resulting in higher breaks at spinning, and uneven yarn. The vibration may be due to worn out flyer pins, unbalanced flyers and worn out top slots in the spindles.
3. **Bad top arms:** Bad top arm gives low draft and makes the rove heavy. These lead to undrafted ends at spinning and also count variations.
4. **Eccentric drafting rollers:** Eccentric top rollers give periodic short-term irregularities in rove leading to long-term variations like long thin places (H and I faults) and long thick places (E, F and G faults) in yarn. The extent of irregularity depends on the position of eccentric

roller in the drafting. For example, the front rollers give 10 cm peak in roving spectrogram, which shall become 2–3.50 m in spinning. If the back roller is eccentric, the peak at roving shall be at around 1 m and give rise to long faults in spinning of 20–35 m.

5. **Loose fitting of bobbins resulting in jumping:** Jumping bobbins disturb the coils on the rove, which give problem while unwinding in spinning creel. This results in uneven stretch and long-term thin places like H faults.

6. **Jumping or vibrating cone drums:** The vibrations and jumps in cone drums give irregular bobbin speed resulting in uneven stretch, leading to count variations at spinning. The jumping of cone drums is due to worn out bearings and shafts.

7. **Deshaped flyers:** The flyers can get deshaped because of the continuous working at a high speed. Also rough handling and allowing the flyer to fall off, especially on other flyers those are running are also the reasons for deshaped flyers. By this we get vibrations and higher breakages (Fig. 4.52).

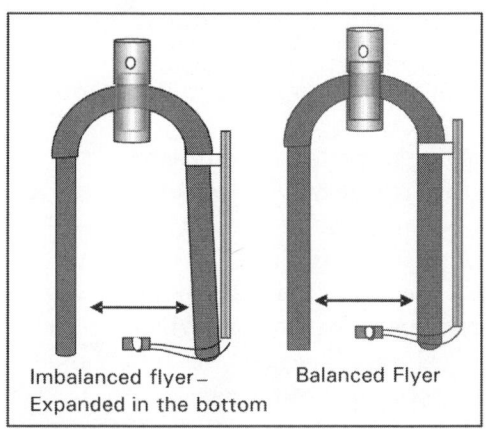

Imbalanced flyer –
Expanded in the bottom

Balanced Flyer

4.52 Deshaped flyers.

8. **Damaged condensers:** The condensers, if damaged, shall restrict the free movement of fibres resulting in chokes up and breaks.

9. **Loose chains driving the creel:** The loose chains give jerky motion to creel resulting in stretching of slivers. This leads to long thin places in spinning.

10. **Loose fitting of gears especially in spindle and bobbin drives:** The loose fitted gears in spindle and bobbin drive give stretch variation in rove and long-term thick and thin places in tarn.

11. **Worn-out clearer clothes:** Worn-out clearer does not clean the cots, and allow the fibres to lap on it leading to breakages.

12. **Loose saddles:** Loose saddles results in tilting of top rollers, thus resulting in uneven pressure and uneven drafts.

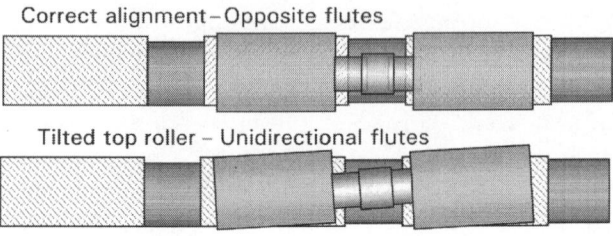

Correct alignment – Opposite flutes

Tilted top roller – Unidirectional flutes

4.53 Loose saddles.

13. **Tilted top rollers:** This is a normal problem in speed frames with a higher gauge as the saddle sizes where the top rollers are held is almost same in all machines. If the helical flutes are in single direction as shown in Fig. 4.53, we get the problem of tilting. By having flutes alternately in opposite direction, this problem can be reduced.

14. **Worn-out needle bearings in fluted rollers:** Worn-out needle bearings give eccentricity to fluted rollers resulting in drafting waves. This leads to higher U% in rove and long-term thick and thin places in yarns.

15. **Worn-out springs:** Worn-out springs in top arms give lower pressure to the top roller. This results in lower and uneven draft giving rise to count variations in spinning.

16. **Worn-out false twisters:** The false twisters are employed to enhance the strength of rove between flyer and the nip of front roller. The false twisters, which are made of fairly soft materials, wear out while working. Vijayakumar[6] suggests replacing false twisters once in two years. He observed that variation in false twister resulted in high count CV%.

4.3.6   Ring frames

1. **Damaged wheels:** Damaged wheels increase end breakage rate in spinning, leading to more piecing. Also it leads to irregular yarns, lower strength due to presence of more weak spots.

2. **Vibrating spindles:** Vibrations in spindles result in higher breaks. They also lead to ring cuts damaging the yarn. The yarn breaks repeatedly at winding because of the ring cuts, and finally the cop gets rejected. The ring cut also blackens the yarn (Fig. 4.54).

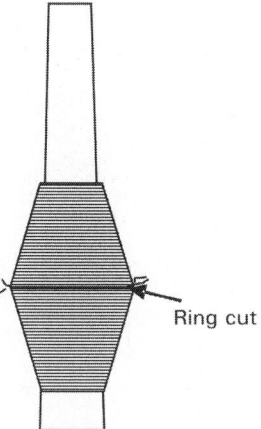

4.54 Vibrating spindles.

3. **Inactive button springs in spindles:** The buttons are provided on plug type spindles to hold the bobbin firm on spindle when the spindles are running at a very high speed. If the button springs are not active, there will not be proper grip and the bobbin might jump, vibrate or run at a slower speed giving raise to ring cuts and weak yarn (Fig. 4.55).

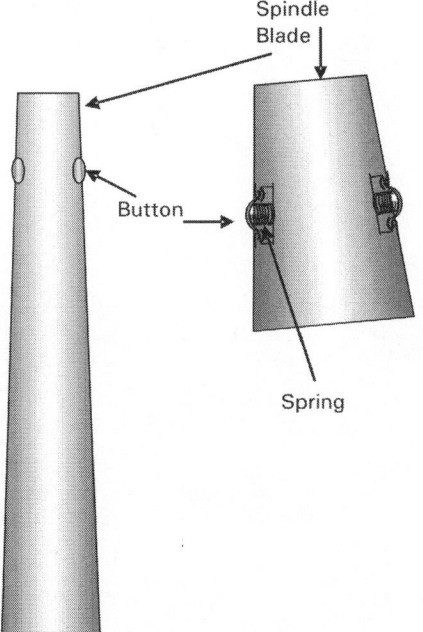

4.55 Inactive button springs.

4. **Fallen buttons in spindles:** The buttons fall off from the spindle if the button holding slot is damaged in the spindle blade. When the

buttons fall off, there will not be proper grip to the bobbins, leading to jumping cops and ring cuts (Fig. 4.56).

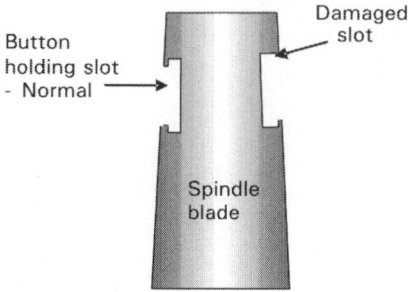

4.56 Spindle blade.

5. **Misalignment of spindles and lappets:** The spindles should be exactly in the centre of the rings and the lappet eye should be centered exactly above the tip of the spindle. Any misalignment results in higher breakages (Fig. 4.57).

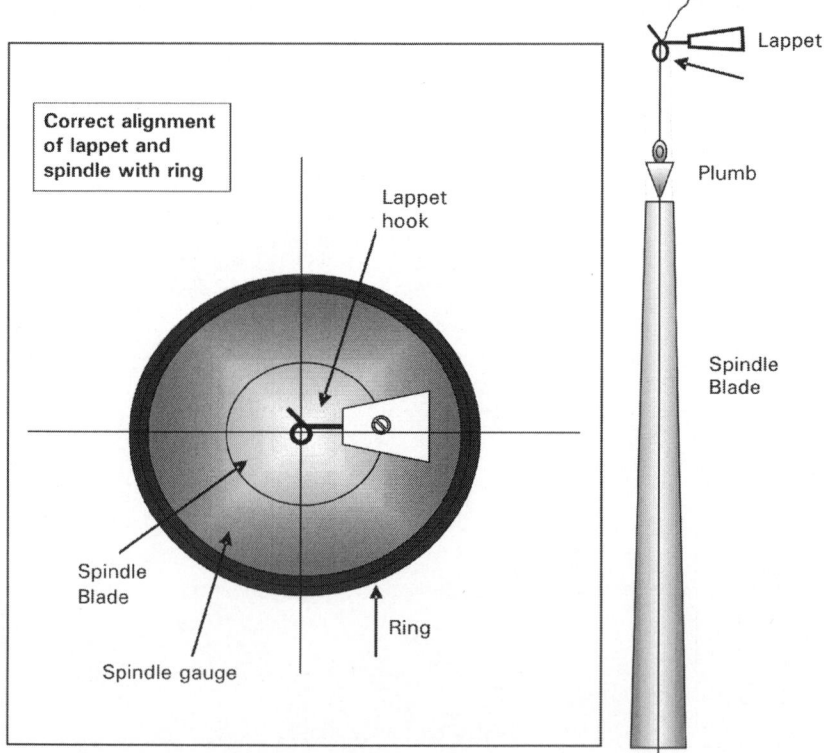

4.57 Alignment of spindles and lappets.

6. **Eccentric drafting rollers:** Eccentric drafting rollers vary the gripping point of fibres and the settings shall be continuously getting changed in a cyclic manner. This gives periodic thick and thin places with a repeat for the circumference of the eccentric roller (Fig. 4.58).

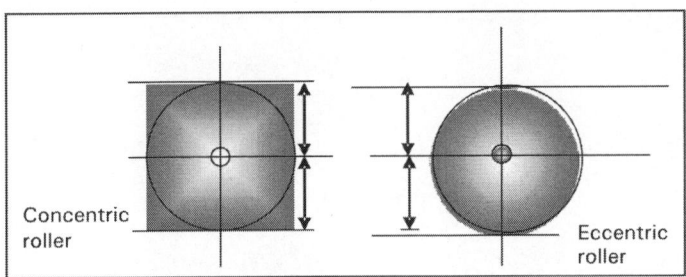

Concentric roller

Eccentric roller

4.58 Eccentric drafting rollers.

7. **Damaged cots:** The quality of the cots directly contributes to the regularity of the yarn produced. Any variations like cut marks, lower diameter, channeling, press marks, tapering, air gaps in mounting, cracks due to uneven hardness of undue exposure to sunlight for a long time, etc., cause short-term variations with uniform repeats (Fig. 4.59).

The condition of cot is very important for getting good quality. Due to long working of cots, the working portion might get worn out and channeled. Channeling also takes place when traverse motion is not working. The air gap develops while mounting the cots. The cots need to be rolled thoroughly with uniform pressure to remove air gaps while mounting.

8. **Bad aprons:** Aprons are provided to control the movement of fibres in uncontrolled zone of drafting. The quality of aprons is very important for getting quality yarns.

Aprons generally come in packed condition with 10 aprons in a bundle. The aprons, which are made of synthetic rubber, are very sensitive to heat. If the aprons are kept in stores for a long time, they become hard and we get hard folding marks. It is therefore suggested to open the package and make all the aprons loose and then store it. Do not keep the aprons, or any rubber material, in folded condition for a long time (Fig. 4.60).

Aprons get channeled because of the non-functioning of traverse guide.

Normally while putting new aprons, we put all endless aprons. However, due to various reasons, the aprons get cut and we need to replace them. It is easy to replace top aprons. In case of bottom aprons, we need to remove the fluted rollers in orders to put endless aprons,

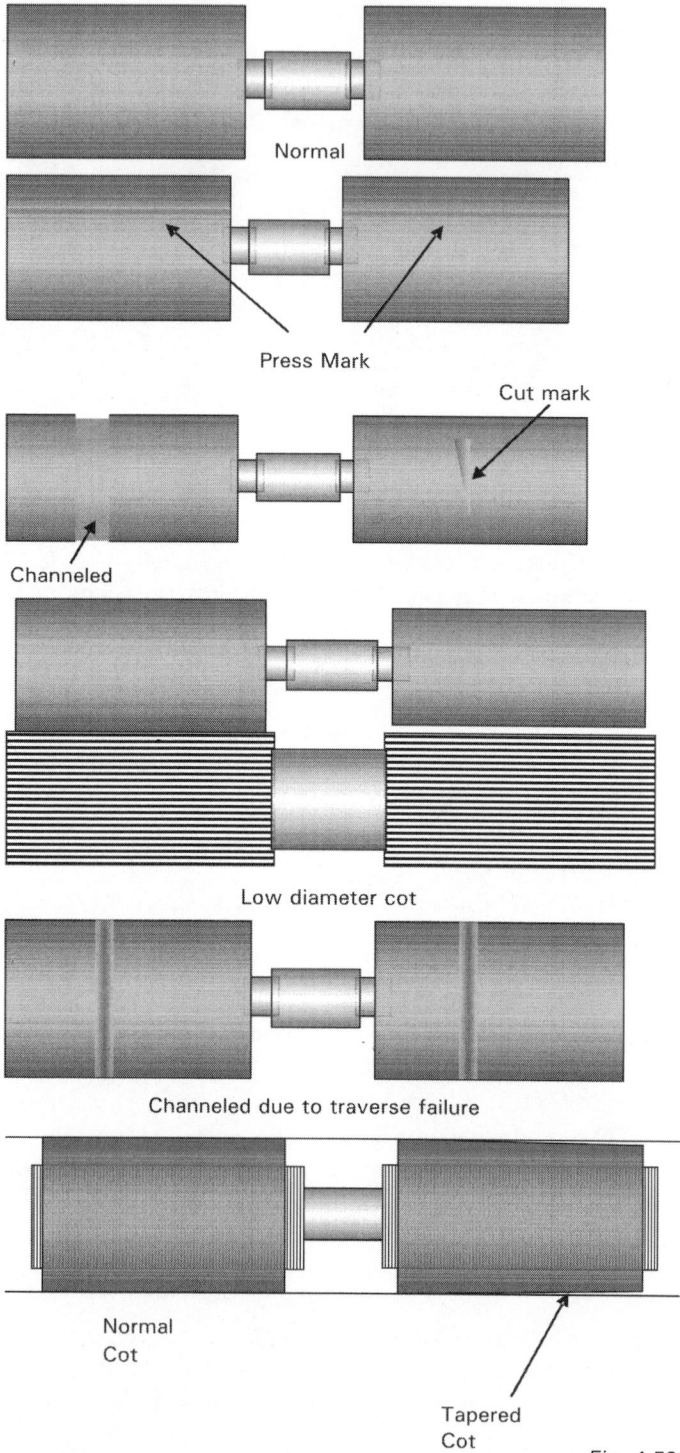

Fig. 4.59 Contiued.

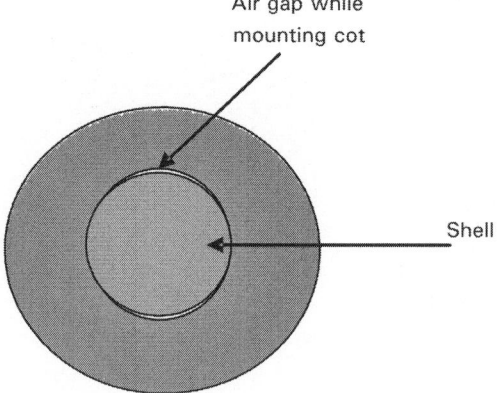

4.59 Condition of cots.

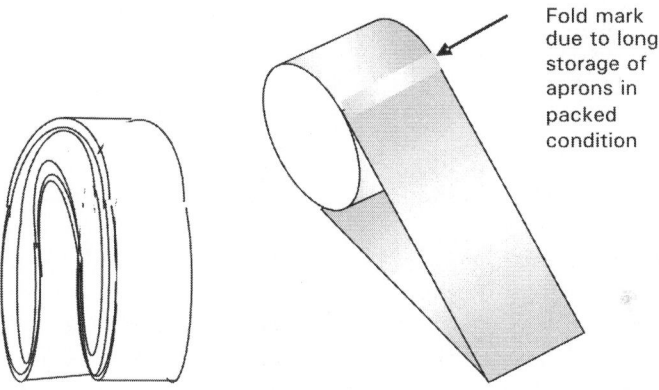

4.60 Aprons folded by supplier.

which is not feasible on a working machine. Therefore, we take open aprons and join them. The joint should be perfect otherwise it will give periodic variations (Figs. 4.61 and 4.62).

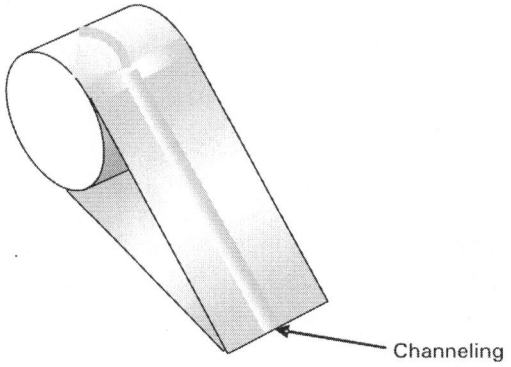

4.61 Bad aprons.

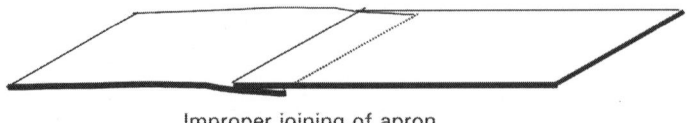

Improper joining of apron

4.62 Improper joining of apron.

Aprons, sometimes, become very thin because of long working. Then also we get poor quality.

One of the main reasons for apron breaking is the oil migration from the roller stands. The bottom aprons in the spindles next to roller stands are the victim of this. Fig. 4.63 shows oil migrating from the roller stand.

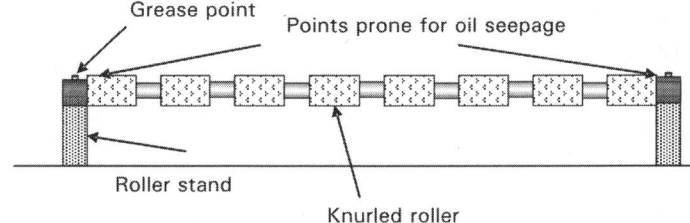

4.63 Oil migration from roller stand.

9. **Worn out rings:** Rings wear due to prolonged working and metal to metal friction between ring and travellers. The wearing out shall be faster in mills where the count changes are frequent, heavy travellers are in use and synthetic spun yarns like polyester are produced. The rings wear out faster by not following the system of replacing the travellers periodically. The rings need to be cleaned with a clean dry cloth before replacing the travellers. This shall ensure removal of carbon accumulation. Worn out rings lead to frequent end breakages and high hairiness.

10. **Burnt travellers:** The traveller burns because of friction with ring. Normally a traveller works for 3–5 days depending on the count and speed. As the travellers start wearing out, we see increase in hairiness and imperfections.

11. **Disturbed/damaged traveller clearers:** The purpose of traveller clearer is to prevent fluff accumulation on traveller, thus increasing the life of traveller and reducing the end breakages. It is found in number of mills that people are not even aware that there is a part like traveller clearer.

12. **Improper alignment of ring rails:** The ring rail alignment is very important. If not properly aligned we end up with low yarn content on a cop and for higher hard waste (Fig. 4.64).

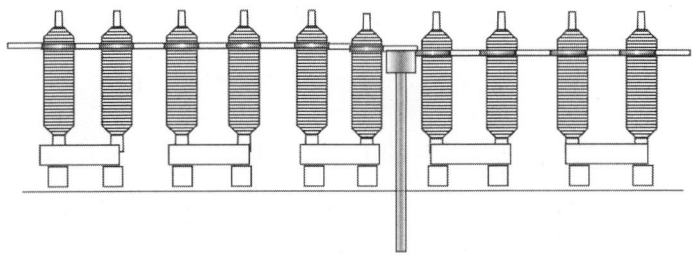

4.64 Ring rail not aligned properly.

13. **Worn out poker bars:** Worn out poker bars result in vibrations and tilting of ring rail while in motion resulting in yarn breaks.

14. **Improper centering of spindles:** If ring diameter is more than 40 mm, ring centering plays a major role. Improper centering results in higher yarn breakages and hairiness variation within the chase will be very high.

15. **Damaged separators:** The separators should be always smooth as the yarn balloon is continuously rubbing on its surface at a very high speed. Any rough surface on separators can increase hairiness of the yarn.

16. **Damaged roving guide:** A damaged roving guide hinders the smooth movement of rove, and introduces stretch. Also it shall be holding some fibres. It might result in breaks or uncontrolled fine count in the yarn.

17. **Serrations in lappet hook:** The lappet hook should always the smooth, as the yarn is continuously rubbing on it. The serrations develop on the lappet hook because the yarn passing through a single path (Fig. 4.65).

Serration in Lappet hook

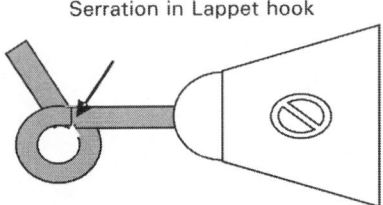

4.65 Serration in lappet hook.

18. **Worn-out covering of clearer cloth:** Clearer rollers are put on the top of front and back top rollers to collect loose fibres coming up while drafting and to prevent lapping. If the cloth is worn out only bare wooden roller will remain that cannot clean the top rollers. Therefore, lapping chances are more (Fig. 4.66).

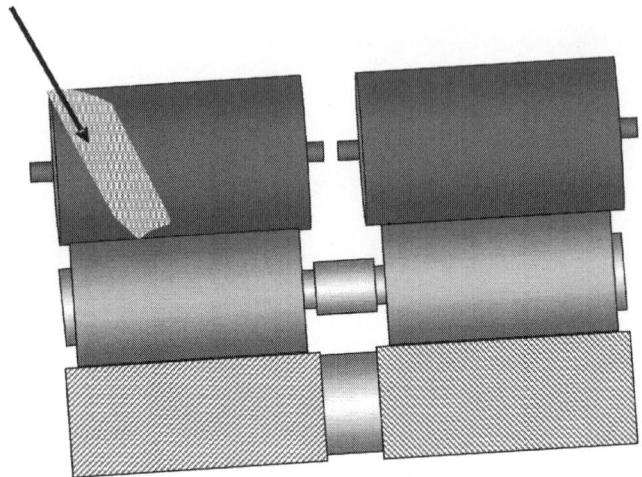

4.66 Torn clearer roller.

19. **Damaged mouth piece of suction unit:** The suction units are provided to suck the fibres from a broken end and prevent multiple breaks. If the moth piece is damaged, it holds the fibres, and suction becomes ineffective. The fibres from broken end are not sucked out and results in multiple breaks (Fig. 4.67).

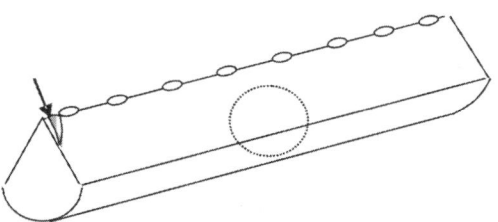

4.67 Torn mouth piece.

20. **Air leakages in the suction duct:** Air leakages in the suction ducts, especially at the joints reduces suction pressure at the mouth piece, which results in multiple breaks.

4.3.7   General

1. Worn out parts: bearings, shafts, links, etc. Any worn out part means inconsistent performance.
2. Loose parts – nuts, bolts, screws, joints, fittings, etc. A loose part disturbs the settings and leads to breakdowns.
3. Eccentric shafts give irregular motion, drafting waves, and results in breakdowns.

4. Vibrations due to loose foundation, worn out shafts, misbalancing of machines, etc., not only increase the breakages and produce poor quality material but also leads to breakdowns.
5. Disturbed settings and timings result in poor quality.
6. Blunt or bent pins, wire points, grids, etc., result in neps, slubs and increased wastes.
7. Bent or scratched covers, screens, perforations, etc., lead to neps and higher imperfections.
8. Broken parts give poor quality.
9. Non-standard parts are likely to damage the costly original parts of the machine. In emergencies we are forced to use non-standard parts as we cannot afford to keep a machine idle. It is suggested to paint the non-standard parts with a different colour so that one can always remember that a not standard part is being run. It should be replaced as early as possible, and kept as a spare for emergency.
10. Air leakages create lower pressure in drafting, malfunctioning of splicers and automatic mechanisms like feed regulations doors in blow room, doffing of the laps, cans, etc.
11. Water contamination in the compressed air reduces the supply of compressed air to resulting in failure of critical mechanisms of feed control, doffing, splicing, drafting, etc. The water should be removed periodically.

4.4   Technology and parameters adapted

Technology and the parameters adapted play a vital role in deciding the quality of the yarn spun. Poor quality can be a result of the following:

(a) Adapting a technology verifying its appropriateness for the product and the activity, but just because it is latest.
(b) Not updating the technology for the requirement of the product.
(c) Not understanding the available technology, and hence not optimizing the parameters as needed.

Let us discuss some of the effects of different parameters and technology on the yarn quality.

4.4.1   Blow room

The operations in blow room need to be controlled depending on the quality characteristics desired in yarn. Some of the important yarn quality characteristics influenced by blow room performance either directly or indirectly are as follows:

   (i) Evenness – by level of short fibres generated (increase in fibre rupture)

  (ii) Imperfections – by level of short fibres generated, degree of cleaning and opening and neps generated in blow room

 (iii) Neps – by additional neps generated in blow room

 (iv) Hairiness – by level of short fibres generated

  (v) Consistency in count – by delivery of uniform product either lap or the feed to cards

 (vi) Consistency in property – by degree of mixing achieved

(vii) Cleanliness – by degree of cleaning achieved and possible rupture of large trash particles

1. **Improper selection of cleaning and beating points:** Selection of cleaning and beating points, and their sequence is very important or else we end up with either harsh beating where fibres needed gentle beating or inadequate beating where high beatings were needed. Dr. R. Chattopadhyay[7] groups the opening and cleaning activities in a blow room in four categories like plucking, interaction between tufts and opposing spikes, impact by strikers on tufts in free flight and teasing action by needle or pins or saw tooth elements on tufts in nipped state. Out of these four, the first one is mild in nature whereas the second and third are moderate and the fourth is most intensive. Machines need to be arranged so that the most intensive openers are at the later stage since the opening becomes progressively difficult with decreased tuft size.

2. **Improper selection of speeds, settings and ratio of beater and fan speeds:** Speeds and settings of blow room machines play a very important role in deciding the amount of opening and hence cleaning that can be achieved by a machine. Higher beater speeds and closer settings normally give better cleaning, but if used carelessly, it can lead to fibre damage, nep formation and stringiness of cotton. One needs to understand the technology and decide the parameters. Dr. R. Chattopadhyay[7] explains the ways of changing intensity of each type of opening machines as follows:

   (a) **Plucking:** Reduction in the depth of penetration. As this shall have a negative effect on production, the speed of plucking needs to be increased to maintain productivity.

   (b) **Interaction between opposing spikes:** The intensity can be changed by altering speed and setting between interacting surfaces, varying spikes density and the throughput rate.

   (c) **Impact:** The intensity can be varied in this type of opening and beating machines by altering the velocity of strikers (speed of rotation), density of strikers, closeness of line of action of strikers and material grip point, i.e., setting between feed roller nip and the beaters, speed of suction fan and throughput rate.

(d) **Teasing:** The intensity of cleaning and opening can be altered by changing the speed of beaters (needle or saw tooth roller), density and angle of inclination of needles or saw tooth, setting between feed roller nip and the beating point and the throughput rate.

Beater types and specification should be selected properly based on the positions of the beater and the type of raw material (fibre micronaire and trash percentage).

3. **Inadequate controls:** Inadequate controls of feed, delivery, tuft size and production rate results in unopened tufts, jamming of machines, uneven feed to cards and stoppages. Dr. Chattopadhyay suggests manipulation of parameters of those machines which treat fibres gently rather than impact and teasing type machines to avoid damages. Once these machines, which are normally kept in the beginning of a line, are optimized, then we need to move to other machines. A reduction in throughput rate (production rate) and reduction in the thickness of feed improves opening capability in all machines. Changing the density of needles, pins, spikes, saw tooth, type of beaters, etc., are difficult but altering the speeds and settings are the easy. Therefore, we should first try to optimize these conditions before thinking of changing the machine configuration. Controlling the humidity conditions is also very important as damp weather gives poor cleaning and opening.

4. **Longer lengths of pipes/ducts and excessive bends in the material passage:** The longer length of pipes directly increases the neps level because of the curling of fibres. Also each bend is a potential place for generation of neps.

5. **Improper control of air currents:** If air current are not controlled it may result in:
    (a) back firing
    (b) uneven accumulation on condensers
    (c) inadequate removal of micro-dust
    (d) slow movement of materials over beaters
    (e) inadequate beating because of very fast movement

Increase in neps after blow room process should be less than 80% (i.e., 180% of raw cotton nep). If the nep increase is more, then beater speeds should be reduced instead of feed roller to beater setting. If the trash percentage in cotton is less and the neps are more in the sliver, number of beating points can be reduced.

Variation in feed roller speed should be as low as possible especially in the feeding machine. Vijayakumar[6] suggests keeping the material pressure in the ducts as high as possible to reduce feeding variation to the cards. The feed rollers in the chute should work continuously without more speed variation if pressure filling concept is used (i.e., balancing of the chute should be done properly). For others, the feed roller should work at the maximum speed for a longer time.

Material density between different chutes should be same. The difference should not be more than 7%. Vijayakumar insists that the difference in duct pressure should not be more than 40 Pascal in chute feed system. Air loss should be avoided in the chute feed system, to reduce the fan speed and material velocity.

Blow room feeding should be set in such a way that the draft in cards is same for all the cards and the variation in feed density is as low as possible.

4.4.2   Carding

1. **Wrong selection of feed and hank organization:** The feed and hank organization has direct link to the fibre length and fineness, and the final count of yarn to be spun. The feed hank and the drafts also depend on the condition of the carding machine. A very low draft results in lesser individualization of fibres. Normally a draft of 100 ± 10 is found to give good results when laps are fed. In case of chute fed cards, still higher drafts can be adopted.

2. **Wrong selection of wire points for the material and hank:** The wire points selected depend on the fibre fineness, the production rate and the type of fibre. Fine cottons need higher points per square inch. The polyesters need coarser wires. With higher points per square inch, the quality of carding shall be good, but the chance of wire damage is more with coarser micronaire and coarse hanks. Also the heavy trash gets loaded in the fine wires easily resulting in cylinder loading. In case of polyester fibres we notice cylinder loading with fine wires and higher density.

3. **Wrong selection of speeds and settings of different parts:** Setting between cylinder and flat tops should be as close as possible, depending upon the variation between cylinder and flat tops. Care should be taken so that wires do not touch each other. The cylinder speed should be decided considering the micronaire value of cotton. Vijayakumar[6] suggests that if the micronaire is lower than 3.5, the cylinder speed should be around 350 rpm. If the micronaire is between 3.5 and 4.0, it can be around 450 rpm. If the micronaire is more than 4.0, it can be around 500 rpm. He further suggests that lower the

micronaire, lower the licker-in speed. It should range from 800 to 1150 rpm depending upon the micronaire and production rate.

4. **Wrong selection of feed plates for the fibre in use:** Short nose feed plates are to be used for cotton up to 34 mm length and long nose feed plates for staple fibres like viscose and polyester where the staple length is more. However, it is seen that in a number of cases, due to urgency, the cards are allotted for different fibres, and only the speeds like licker-in speeds, flat speeds and doffer speeds are changed. The wire points and the feed plates are not changed. By this we cannot expect good quality of web as fibres are not fed to licker in properly.

5. **Improper control points for the activity:** Unless we have identified proper control points and monitor them, it is not possible to get the required quality. Depending on the quality requirements, we need to decide the control points and check points.

6. **Inadequate control of humidity and temperature:** Humidity plays a very important role in working of cards and the quality of web. A dry weather results in too much fly generation due to static charges and higher breakages. A higher humidity results in sagging of web.

7. **Improper setting of auto-levellers:** Card auto-levellers should be set properly. Nominal draft should be correct. Draft deviation should not be more than 5% during normal working.

8. **Stoppages:** Card stoppages should be as low as possible. Any stoppage in a card means a break in the sliver. There shall be undue thick or thin sliver while stopping and restarting a card. Each stoppage and restarting is a potential generator of slubs in the yarn.

9. **Slow-speed working:** Vijayakumar[6] suggests avoiding of slow-speed working of cards and removing the slivers produced during slow speed. If the card is designed only for slow speed, there is no problem in running at slow speed, and we get good quality. However, if the card runs for sometimes at a slow speed and again at a fast speed, we get difference in quality. There shall be variations in the hank of sliver and also the fibre opening.

4.4.3 Draw frames

1. **Improper adjustment of break draft and final draft combination:** The combination of break draft and final draft is very important to get correct quality. It mainly depends on the fibre fineness, fibre length and the total mass of fibres being drafted. Studies need to be conducted in each mill to get the best combination.

2. **Improper settings for the hank and material being worked:** The settings not only depend on the staple length, but also on the bulk of the materials. If the bulk is more, i.e., the quantity of fibres getting

drafted is more we might need to keep slightly wider setting compared to what we keep for finer mass.

3. **Improper settings of the autoleveller:** Autolevellers need to be set depending on the fibres used and the hank. It should be validated by conducting sliver test.

4. **Hooks formations because of the type of trumpets used:** In normal trumpets, as the drafted web is converging through the small opening of the trumpet, the air in between fibres has to escape and come out. As there is no sufficient place for the air to come out, it has to take a reverse path, resulting in the hooking of the fibres, which were earlier made straight by the drafting action (see Fig. 4.68). Hence the purpose of draw frame is defeated. By drilling two are three holes in line with the direction of movement of air, we can prevent the air taking a reverse path. By this we can avoid formation of hooks, and can adapt a still smaller trumpet for the same hank, and get better quality of sliver. The trials conducted by the authour on Rieter DO 2S draw frames showed a reduction in U% by about 0.5. The theory holds good for all machines, not only draw frames; where the materials are condensed and made to pass through a trumpet.

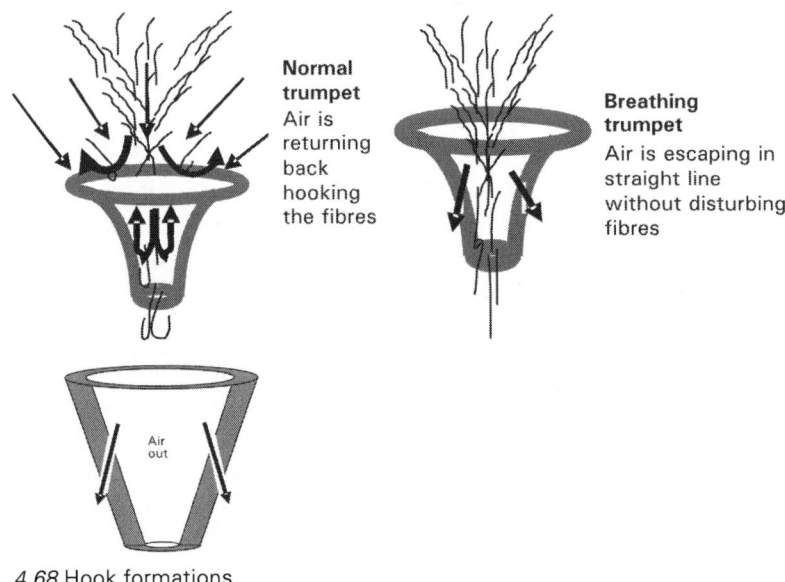

4.68 Hook formations.

4.4.4   Combers

While designing lap preparation, total draft, fibre parallelization, number of doublings, lap weight, etc., should be decided properly. This can be done by conducting trials as the parameters depend on various factors like the machine, the working environment, the staple length and the short fibre contents.

1. **Improper selection of half lap:** It was normal practice earlier to have 16 combs in a half lap. The first comb used to be the coarsest, and the second one with slightly finer needles and higher density. The needle density increased gradually, with the last comb having the finest needles and highest density. However, with high-speed combing, the damages to the half lap was more, especially in the last 3–4 rows. The Nitto Unicomb was developed to avoid this problem, with one sheet having needles in a twill shape. There was some reservation on the quality, as Unicomb was designed mainly to get higher production and longer life for combs. Steadler developed combs in four modular sets, each with four lines of needles for easy replacement and to get a better quality. One needs to understand the requirements and decide on the half lap (Figs. 4.69 and 4.70).

2. **Improper selection of trumpets:** The trumpets used should match the hank of sliver being produced. A finer trumpet leads to more chocking ups and stoppages, whereas a larger trumpet falls forward because of low tension and results in stoppages (Fig. 4.71).

3. **Inadequate control of humidity and temperature:** Comber is very sensitive to humidity and temperature compared to other machines in spinning section. It is highly essential to maintain the required humidity and temperature at all the time to get good working and good quality.

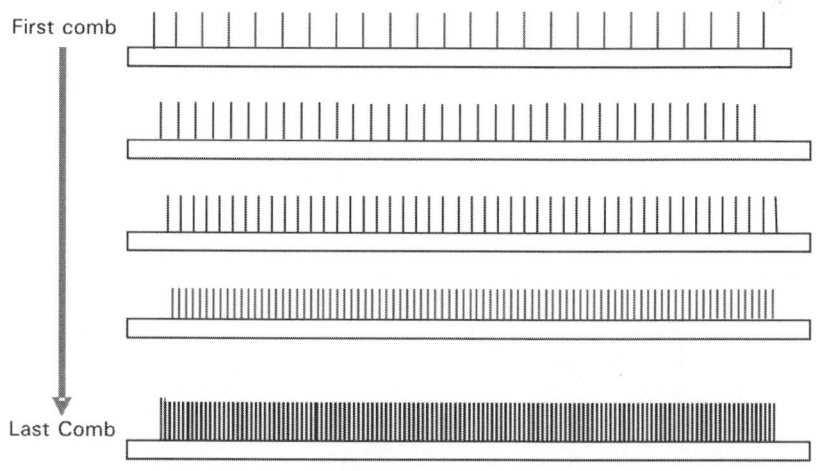

4.69 Combers.

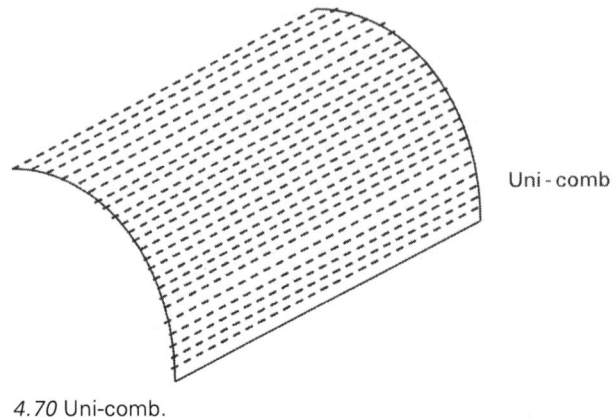

Uni - comb

4.70 Uni-comb.

Correct position of trumpet

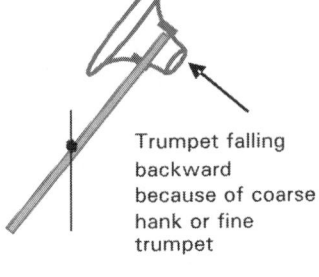

Trumpet falling forward because of fine hank or wider trumpet

Trumpet falling backward because of coarse hank or fine trumpet

4.71 Improper selection of trumpets.

4. **Improper selection of feed for the noil% extracted:** There are two types of feed available in combers, viz., forward feed and backward feed. The forward feed is good for Noil% up to 16%, and backward feed for higher noil extraction.

Vijayakumar[6] gives the following tips for getting good quality from combers:

- In lap preparation, total draft, fibre parallelization, number of doublings, lap weight, etc., should be decided properly (based on trial).
- Higher the lap weight (grams/meter) lower the quality. It depends upon the type of comber and the fibre micronaire.
- If fine micronaire is used, lap weight can be reduced to improve the combing efficiency, if coarse micronaire is used, lap weight can be increased.
- If fibre parallelization is too much, lap sheets sticking to each other

will be more (it will happen if the micronaire is very low also). If the lap sheets are sticking to each other, the total draft between carding and comber should be reduced.

- If the draft is less, fibre parallelization will be less, hence loss of long fibres in the noil will be more.
- Top comb penetration should be maximum possible for better yarn quality. But care should be taken so that top comb will not get damaged. Damaged top comb will affect the yarn quality very badly.
- Setting between Unicomb and top nipper should be same and it should be around 0.4–0.5 mm.
- Feed weight is approximately 50–58 g for combers like E7/4 and is 65–75 g for combers like E62 or E7/6.
- Lower the feed length, better the yarn quality. Trials to be conducted with different feed lengths and it should be decided based on quality and production requirement.
- Required waste should be removed with the lowest detaching distance setting.
- For cottons with micronaire up to 3.5, top comb should have 30 needles/cm and for cottons with more than 3.8 micronaire, the top comb should have 26 needles/cm.
- Trials to be conducted to standardize the waste percentage.
- Piecing wave should be as low as possible. Piecing index should be decided based upon cotton length and feed length.
- Spectrograms should be attended to get lower sliver U%.
- Head-to-head waste percentage should be as low as possible.
- Variation in waste percentage between combers should be as low as possible (range less than 1.5%).
- If cotton with low maturity coefficient is used, it is better to remove more noil to avoid dyeing variation problem.

4.4.5   Speed frames

1. **The drafting system not suitable for the fibres and hank:** There are various points in a drafting system which needs to be suitable for the materials and the hank being processed. For example:
   (a)   Cradle size
   (b)   The apron bridge bar
   (c)   The settings and top arm pressure combination
   The cradle size depends on the staple length. Normally for cottons 36 mm cradles are used. The cradles are also available in 44 and 51 mm gauge. The cradles are named as short cradle, medium cradle and long cradle. Depending on the cradle size, the apron bridge bars also need to be changed in some drafting systems like WST UTM

620. If the apron bridge bar is supported by the front roller slide, then there is no need to change Apron Bridge bar when a cradle is changed. In systems where the apron bridge bar is supported by the middle roller slide, then there is a need to change the apron bridge bar also (see Figs. 4.72 and 4.73).

The alignment of top rollers depends on the type of top arm pressuring system. The alignment should be such that we get maximum pressure on the top rollers. The maximum setting possible on a top arm drafting system should be studied before accepting a long fibre. Normally all top arms are designed for cottons, and if we need to run long staples like 51 or 60 mm, the system shall not be suitable unless otherwise it is designed for the same.

2. **Improper selection of break draft and main draft for the fibres and the hank:** The combination of break draft and main draft is very important to get uniformity. One needs to study this by conducting trials and fix the parameters. Break draft normally adopted ranges from 1.06 to 1.28 depending on the type of drafting, total draft, the fibre length and fineness. Roving hank should be decided in such a way that the ring frame draft is around 16–38 for different counts.

3. **The flyer size/type not suitable for the hank and the material in use:** The flyer size needs to be selected considering the bulk of the material. We need wider passage for polyesters compared to cotton for the same hank of rove because of the density difference of fibres.

4. **The bobbin lift and diameters not suitable for the hank of material produced:** The bobbin lift and diameters directly influence the tension of rove. Roving tension should be as low and as uniform as possible. Higher the roving tension, higher shall be the count CV% and also higher thin places. Undue low tension also creates problems like soft bobbins and breaks in ring frame creel. Density of all roving bobbins should be same. Higher the variation, higher the count CV%.

5. **Improper selection of condensers and spacers:** Condensers and spacers are decided considering the hank of sliver and the rove, the type of fibres used and the draft employed. The machinery manufacturers give a chart of various condensers available with them and the range in which they can be used. It is better to use floating condenser in the front zone to reduce hairiness and the diameter of the roving. The workers need to be trained to put back the floating condenser after attending breaks. Care should be taken to increase the front zone setting by 2 mm while using floating condensers.

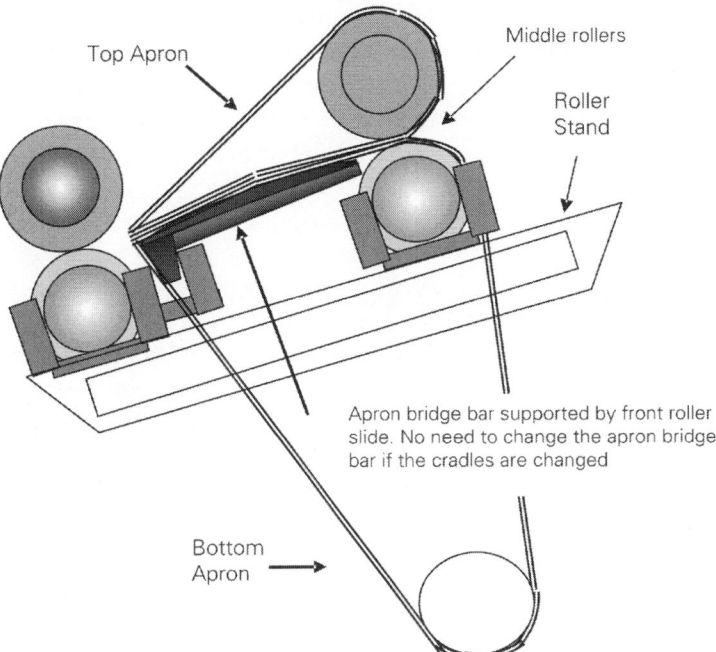

4.72 Apron bridge bar.

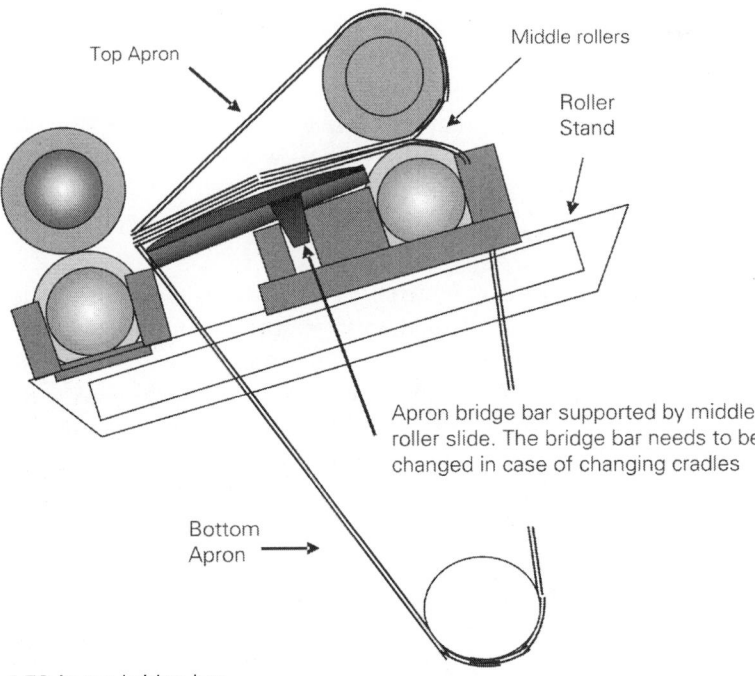

4.73 Apron bridge bar.

6. **The twist multiplier not compatible with fibre length, hank and the drafting system at ring frame:** The twist multiplier selected should be such that it would not allow the rove to stretch, and at the same time does not give undrafted ends at Ring Frames. A higher Twist Multiplier reduces stretch in Ring frame creel, resulting in lower thin places, viz., Classimat "H1" faults.

    If the speed frame bobbins are of larger size, say 1.0 kg and above, it requires more strength for unwinding at Ring Frame Creel. Therefore, slightly higher TM is needed compared to that used for old inters and roving where the bobbins were of 250, 600 g, etc. If the ring frame drafting is good; we can give higher TM at speed frame. If the ring frame drafting does not have capacity to draft, it is better to keep slightly lower TM at speed frame and avoid undrafted ends.

7. **Twist multiplier not suitable for the season:** The hardness of roving bobbin is influenced by the humidity conditions. If the RH% is more, the bobbins become hard, and drafting becomes difficult in ring frames. Therefore, in rainy season, we need to reduce the TM at speed frames. In dry season, the roving bobbin becomes soft, and hence we need higher TM.

8. **Speeds not suitable:** The spindle speed at speed frame has a great impact on the quality. At higher speeds, the tension on rove shall be high that lead to hairiness and fly liberation, and ultimately breakages. Vijayakumar[6] suggests running at less than 1000 rpm in case single speed for flyer is used. When the speed frame bobbin is full, flyer speed should be less than 1000 rpm. Otherwise surface cuts will increase and thin places also will increase.

4.4.6   Ring frames

1. **The drafting system not suitable for the fibre, count and the hank organization:** The drafting system should be suitable for the fibres being used. The diameter of the rollers, the cradle size, provision for setting as per the fibre length, the capacity of the top arms to apply pressure are to be engineered to the fibre. For example in some drafting systems the maximum length of the slots in the roller stand for setting is 135 mm from front to back roller. When normal cottons are used (medium staple) as setting of 43 + 60 is kept. Hence there is no problem. If we need to use longer fibres like 51 or 60 mm, these roller stands may become inadequate. In some old drafting like Casablanca A-500, there was no provision for setting. The front roller diameters used to be 25 mm for cotton, which is not adequate while running long staple fibres.

2. **Improper selection of break draft and main draft combination for the hank and count:** Selection of combination of break draft and main drafts are very important for getting uniform yarn. However, it is seen in numerous cases, that people do not touch the break draft but alter only the front draft. With in a range of count and total draft, it is okay; but when the change is significant, we need to study and set the drafts.

3. **Improper selection of empty bobbins:** The quality of empty bobbins is very important to get good working on the ring frames. Improper fitting leads to jumping or vibrations, creating higher breaks and improper winding leading to higher breaks at cone winding. Sometimes the yarns are conditioned on cop stage using autoclaves or high pressure steam. The material used for empty bobbins should be suitable for this purpose. The spinner should specify this to the bobbin supplier while placing orders, or else, he might have to face the problem of improper fitting of empties after one or two uses.

4. **Improper settings:** The settings are very important for getting the good quality which depends on the staple length. It is always advisable to have a lower break draft with a wider setting in the back zone. For example, earlier, in top arm drafting like SKF-PK-211, a break draft of 1.3 with back roller setting of 52 mm was common for all cottons, but now a Break draft of 1.14 with back zone setting of 60 mm is common. The gap between front top roller and apron nip should be as low as possible (around 0.5–1 mm). If it is more imperfections will be high.

5. **Improper selection of spacers:** The spacers play an important role in controlling the fibre flow in the front zone of the drafting. The size of the spacer depends on the count being spun. We need small spacers for fine count and wider spacer for fine count. In case of staple fibres like polyester and viscose, normally wide spacers are used compared to the same hank and count of cotton (Fig. 4.74).

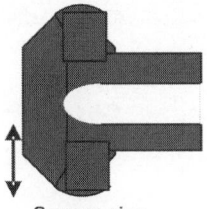

Spacer size

4.74 Selection of spaces.

6. **The hardness of the cots not suitable for the fibres, hank, the counts spun and the speed:** To get good evenness of yarn softer cots are recommended in front. Normally 65° shore hardness for front

rollers and 80° for back rollers is common. In case of polyester and viscose, although 65° hardness gives better results, the life of the cots reduce significantly. This becomes unviable. Hence 85° and 90° hardness cots are used.

7. **Improper synthetic rubber:** The quality of synthetic rubber should be suitable for the working conditions like speed, heat generation in the spinning shed, the humidity condition and the pressures. A wrong selection might lead to higher breakages, lapping problems, increase in imperfections, etc. It is advisable to discuss with the cot and apron manufacturer and select suitable rubber depending on the fibre, working condition and the speed.

8. **The lift not suitable for the count:** A smaller lift gives better working whereas a longer lift gives better cop content. In case of coarse counts, the number of doffs become very high with shorter lifts, and hence the number of bobbin changes and the splices increase leading to a poor performance. In case of fine counts, using a longer lift gives poor working in spinning, and the power consumption shall be high. The lift used should therefore be suitable for the count.

9. **The rings and traveller not suitable for the fibre, count and the speed:** There are number of factors contributing for quality and productivity in ring frames that are related directly to ring and traveller. They are as follows:
   (a)   Ring diameter
   (b)   The flange
   (c)   The type of ring
   (d)   The surface finish
   (e)   The type and profile of the traveller
   (f)   Compatibility of rings and travellers

   Geocities.com[6] attributes the limit to productivity in the ring spinning machine to the interdependence of the traveller with the ring and yarn. It is very important for the technologist to understand this and act on them to optimize the yarn production.

   The following factors should be considered:
   • Materials of the ring traveller
   • Surface characteristics
   • The forms of both elements (ring and traveller)
   • Wear resistance
   • Smoothness of running
   • Running-in conditions
   • Fibre lubrication

10. **Travellers:** It is better to use lighter travellers than using heavy travellers. Enough trials should be taken, because traveller size depends upon, speed, micronaire, humidity condition, count, the

profile of ring, ring diameter, etc. The ring travellers should be changed before 1.5% of travellers burn out. Whenever there is a multiple break, ring travellers should be changed. At any point of time, fluff accumulation on travellers should be less. Ring traveller setting should be close enough to remove the waste accumulation but at the same time it should not disturb the traveller running

11. **The lappet eye not suitable for the fibres and count spun:** The lappet eyes play an important role in deciding the yarn hairiness. The surface finish of the lappet eye, where the yarn is continuously getting rubbed is to be specifically designed depending on the abrasion resistance of the fibres. Other factor in the lappet eye is the diameter of the lappet eye. We need wider eyes for coarser count and smaller eyes for fine count.

12. **The condensers not suitable for the hank and count:** The condensers control the fibres from getting spread during drafting. The size of the condensers depends on the hank of material fed and the count being spun. Similarly the spacers also depend on the count being spun.

    The manufacturers of condensers and spacers need to follow standard colour codification. There are instances that the mills try to develop vendors for cutting the costs. The mills give samples to the vendors for development, but fail to explain the critical measurements. There were incidences of vendors producing the same size of spaces and supplying them with different colours, as the order specified white spacers, red spacers and black spacers, and a sample of black spacer was given. The vendor produced all as per the size of black spacer, but used different colours. Unfortunately, in a number of mills, even the shop floor technicians do not know the method of measuring the spacer dimensions. There is a need to educate the vendors and the users on the standard dimensions and the method of verification.

13. **Improper settings of overhead travelling cleaners:** The travelling clearers are meant for sucking the floating fibres and prevent accumulation of fluff on bobbins and machine parts. Wrong settings lead to blowing of dirt on yarns or disturbing the surface of rove.

14. **Separators not suitable for the lift and speed:** The separators should match with the lift in the dimensions, and to speed in the materials used for separators. A smaller separator for a lift can cause touching of balloons leading to higher breaks. If the material used for making separators are not compatible to the material being spun and the speed, we get problems like higher hairiness and serrations on the separators.

15. **The combination of winding and binding coils not suitable:** The combination of winding and binding coils, the chase length and the

taper are decided on the type of winding machine adapted. The factors are as follows:

(a) Coils per inch and chase length – Higher coils per inch can lead to sloughing off, whereas low coils per inch leads to a lean bobbin.

(b) Motion of ring rail – There are two possibilities, i.e., fast up slow down or slow up fast down. Normally, fast up and slow down is preferred by spinners to have lower breakages. A slow up and fast down is preferred to have easy unwinding at winding.

(c) Ratio of winding to binding coils – The ratio of winding to binding coils needs to be studied and decided depending on the working at winding.

(d) The speed of ring rail movement – a higher speed results in a tight bobbin but breaks shall be high, whereas a slow speed shall result in soft bobbins which might lead to sloughing off while unwinding at high speeds.

4.4.7 Winding

1. **Selection of cone angle to the speed of the machine:** We get damages in the cone tip in case of cones with higher degree of taper like 9°15' compared to 5°57' cones when the speed of winding is high. This problem is more with synthetic staple fibres.

2. **Improper selection of empty cones:** The empty cone quality plays a very important role in getting good winding and cone quality. The cone should fit firmly and should not vibrate or move to and fro. Improper fitting of the cone is one of the main reasons for stitches in the cone. Care should also be taken in specifying the material used for making cones. We have paper cones, wooden cones, plastic cones and stainless steel cones. The paper cones are normally used when the yarn is sold as the finished product of the mill. The strength of paper cone is very important, especially when the winding is at a high speed. Some mills have a practice of reusing the cones; in such cases, they should verify the stability of the cone before using. The wooden cones are normally used when yarn are intended for in-house warping. The plastic cones are also intended for internal use. We might use plain plastic cones for the yarns going for warping, whereas perforated cones in case of the yarns going for wet processing. Stainless steel cones are normally of perforated type and used for yarns going for wet processing in HTHP package dyeing machines. Sometimes the cones go for steam conditioning. The paper or the plastic used should be suitable for steam conditioning, or else the cones become de-shaped.

4.5   Management systems adapted

The management systems adapted play a major role in getting the required quality of materials in time at the estimated cost. Following are some of the mistakes normally done:

(a) Following a system designed in the past that is not suitable today
(b) Following a system followed at other mill not suitable for us
(c) Poor house keeping
(d) Improper/Inadequate communication
(e) Not maintaining log books and records in a simple way

Following are some of the examples of the effect of management systems adapted on the yarn quality:

1. **Not clear about the quality requirements of each process:** Each process should consider their next process as a customer and identify the needs of the customer and design the processes to suit them. Also one should have clarity that the processes designed address the requirement of final customer also. But in a number of cases it is seen that the mills are following some norms ad-hoc because it was recommended by some research association or a consultant. There is no relation to the external customer requirement. Because of this they might produce materials with over specifications resulting in a loss to the company or produce inferior materials resulting in rejections. Each mill need to study the effect of different process parameters in their mill to the final quality and decide the parameters. It is needed to fix targets for each process and monitor the achievement. The target needs to consider the capability of the system, best running environment and intricacies of technology.

2. **Not clear about the capability of the systems:** In a number of cases it is seen that the top management demand the results without realizing that the system installed was not capable of producing that result. For example, the blow rooms are designed to give a cleaning efficiency of up to 70%. This is the maximum figure one can achieve which depends on the trash content and the type of trash in the cottons. If management insists on higher cleaning efficiency, the technicians try closer settings and higher beater speeds resulting in fibre rupture, excess nep generation, extracting of more wastes losing good fibres, etc. One must have a clear understanding of the technology in hand and decide on the targets at each process.

3. **Poor interactions with the people working on shop-floor regarding the quality requirements of the customers/products:** The superiors should understand that the level of thinking and understanding between them and their juniors are different. One

cannot expect the juniors to understand everything in one stroke and perform the job to the satisfaction of superiors. It is the responsibility of the superiors to ensure that the juniors have understood the requirements clearly, and are doing the work as per that. The superiors should have proper interactions with their subordinates.

4. **Poor house keeping:** House keeping includes keeping the required materials in the place suitable for doing the works, removing unwanted materials, keeping the surroundings clean all the time, ensuring the protection of materials in process by suitable coverings and stacking systems. In a spinning mill, the fluff liberation and accumulation of fluff at various places is a common scene. These fluffs can get contaminated with the material in process and give a poor quality. Also, excessive fluff in air is injurious to the people working. Concepts of 5-S need to be implemented at all places.

5. **Congested layout of machinery and processes:** A congested layout results in damages to materials, accumulation of fluff on materials and also can lead to accidents including fires.

6. **Keeping excess materials in stock and process as stand by:** In a number of cases we see that the technicians try to keep some material in process as standby to ensure smooth running of their shifts. The material in stock does not add any value; but is a potential source of contamination and mix ups. It is not possible to go on covering all the materials which are kept as stock in process, might be a mixing or sliver, laps or spinning doffs. We need to workout spin plan in such a way that the machines are balanced and the process stock developed is as low as possible. It is also a fact that the fresh materials work better than the materials produced and kept in stock. The concepts of lean management systems and just in time need to be studied and implemented.

7. **Keeping excess spares in the work area for emergency operations:** People do not want any stoppage of machines, and hence keep some spares for emergency. However, we need to study the pattern and keep minimum spares. Excess stocks are likely to be misused or used at a wrong place just because it was available.

8. **Improper maintenance of records:** Not maintaining proper records of the machine history, maintenance, operations, quality and actions taken for corrections and prevention of problems is one of the main reasons for poor quality. The records should explain the activities done and actions taken in a simple way and the people on the spot should develop a habit of referring to records. However, it is seen that the supervisors and spinning masters do not give importance for referring the records from time to time for taking decisions and for problem identification and solving. If all the activities are recorded

properly, it shall be easy to find the reason for a problem and take action to avoid poor quality.

9. **Not reviewing the quality aspects periodically:** Anything that is not reviewed periodically shall loose its importance. The quality aspects are not an exemption. The superiors and the top management should review the quality aspects as a whole, i.e., product parameters, cost of manufacture, rate of production and completing the assignments in time and the customer feedbacks.

10. **Not taking sincere action on the quality control reports:** In a number of cases it is seen that the quality control section is independent of production, and their activity is only to highlight the bad points to the management. The production people get firing for the mistakes highlighted, which camouflages the good works. Therefore, the production people try to hide the facts from quality control people. The quality reports given by the quality control checkers need to be discussed and understood and actions are to be taken at the earliest to prevent bad quality from moving forward.

11. **Not analyzing the reasons for poor quality and taking ad-hoc decisions:** The experience develops bias or mind set, because of which people come to a quick conclusion. However, in a number of cases, the facts might be different than what was experienced earlier, and hence we do mistakes, and our quality is affected. It was a normal practice earlier, about 30 years back, to reduce the twist wheel in ring frame in case of higher breakages. At that time over spinning was the normal practice, and the cottons issued used to be of inferior quality. But now the situations are changed. People are using superior cottons to get a good quality. Hence reducing the twist wheel can create more harm than benefiting the working.

12. **Not respecting the experience, knowledge and intentions of the juniors:** One of the biggest mistakes done normally in a spinning mill is not recognizing the workers as human being with knowledge. The so-called qualified officers insist the workers to follow the instructions given by them, but forget that the workers, those are working all the times on the machines can think and do the works in a better way also. Workers might be illiterate, but are the actual persons working on the spot. They know what is happening, but are unable to explain in a technical language that is presentable. But the real knowledge of a job or situation comes from actual working and not by reading books or attending seminars and classes. The theory of learning explains the extent to which a man can learn and understand is as follows:

    (a) Only by reading books, one can understand up to a maximum of 15%.

(b)   By getting guidance from a guru after reading, the knowledge shall go up to 30%.

(c)   By seeing the work, one can understand up to 50%.

(d)   By doing the work, the level of understanding goes up to 85%.

(e)   By sharing knowledge with others, the knowledge enhances up to 90%.

(f)   By practicing the good systems all the time, analyzing the reasons for failures and implementing corrective actions, the knowledge level goes up and tends to reach 100%.

Therefore, it is essential for the technicians and managers to accept this reality and start respecting the juniors who are on the job, and involve them in improvement efforts to get quality at lower costs and also production of required material in time.

13.   **Running coarse counts and fine counts in near by machines:** The fluff liberation is more while spinning coarse counts. This fluff gets contaminated with the yarns running on adjacent machines. In case fine yarns are contaminated with this fluff, results in objectionable faults or higher imperfections.

14.   **Frequent count changes from fine to coarse and coarse to fine:** The frequent changing of counts not only spoils the quality but also spoils the machines. The machine running on coarse counts need heavy traveller. These travellers create their own path in the rings. When we change over to fine counts, this path obstructs the free movement of traveller resulting in higher breaks. Frequent changing of wheels can damage the key ways and gears. Further the workers find it difficult to adapt themselves to the piecing requirements of these counts. In case of coarse counts, the front roller speed shall be very high and the bulk of drafted strand coming out of the front roller nips also shall be high. In fine counts the front roller speed shall be low, but the spindle speed shall be high. The drafted strand coming out shall be very thin. We need wider opening of the suction tube for coarse counts compared to fine counts. However, with modern machine having single tubes for each spindle, this problem is not faced, whereas in earlier models, on suction tube was catering to eight spindles. The top rollers get channeled while running coarse counts, and need light buffing while changing the machine to spin a fine count. The setting of traveller clearers also needs to be adjusted while changing the counts depending on the size of the travellers. This is a laborious process, and not done religiously while changing the counts, thus resulting in fluff accumulation on travellers leading to breakages.

Frequent shifting and re-erections: In mills where long-term planning is not there, we see frequent shifting and re-erection of machines. This results in wearing out of the parts, misbalancing of the running parts and disturbance in levels and finally leading to breakdowns. The technicians have to spend more time on the dismantling and erection activities and cannot concentrate on maintaining other machines and monitoring the quality.

Normal problems and non-conformities

5.1 Blow room

5.1.1 Low cleaning efficiency

There is a feeling that if the cleaning efficiency is good in blow room, it is doing a good job. The managements are worried about lower cleaning efficiency of blow room lines in recent days and are insisting the technical persons to increase the cleaning efficiency of blow room. However, it should be noted that the cleaning efficiency of a blow room line depends on the trash content in cotton. If the cotton has more trash, the cleaning efficiency in blow room shall be more. In recent days, due to various steps taken in improving the ginning practices, the average trash content is reducing. Therefore, it is natural that the cleaning efficiencies of blow rooms are coming down. It is essential to concentrate on the trash remaining after blow room rather than the cleaning efficiency.

Check the trash content in mixing and the material delivered, by running at least 200 kg of cotton. Check the blow room wastes collected at different beaters and verify whether it is inline with the norms based on trash%, and also check the lint content in wastes taken out. Increase the wastes if the lint in the wastes is negligible or nil.

If the beater speeds are lesser than required, we get lower cleaning. Therefore, check the beater speeds before taking further steps. Also the ratio of fan speed to beater speed is another important factor to be considered. If the fan speed is more, the cotton misses the beating and the cleaning shall be less. If the fan speed is low, the movement of cotton shall be slow and they get more beatings, and also might result in jamming.

Check the sharpness of the beaters and the beater settings. Correct them if needed. Increase the space between the grid bars so that heavy trash particles can fall down. If the grid bar setting is less, there are chances of blocking of the gaps in the grid area. By choosing a suitable fan speed, even with wider grid bar setting, there shall be no chances of good fibre falling down along with the trash and we can get good cleaning. Close slightly the air inlets under the grid bars towards the cotton entry side, and open those on the delivery side.

If the cottons are moving very fast and because of that cleaning is found less, reduce the fan speed following the beaters by 100–200 rpm.

If the grip of the feed roller is less, we shall get low cleaning efficiency. Therefore, check for the grip.

Check the synchronisation of the machine working. The blending bale openers should work for 80–85% of the time of working of the final machine. If the back machines are running continuously, it is an indication of lower feed to the next machine, but it cannot reduce cleaning efficiency; on the contrary if the back machines are stopping for more time, there is a scope to improve cleaning efficiency by closing down the settings.

If there is a back draught because of not cleaning the wastes under the machines, the cleaning efficiency shall come down. Therefore, ensure that the wastes are removed from time to time.

The cleaning efficiency shall be low if the tufts are not opened properly. Therefore, check the tuft size being fed to the blow room. The problem is greatly reduced by the introduction of bale pluckers which pick very small tufts of cottons of 20–25 mg in place of very big lumps of over 50 g in case of normal hand opening and around 25 g in case of mixing bale openers.

5.1.2   High nep generation and fibre rupture

Nep generation and fibre rupture are mainly due to excessive beatings, rubbing of fibres on a rough surface, entangling of fibres, jamming of fibres in the machines and damp fibres fed in blow room.

Check for blunt beaters, burrs in grid bars, bent nails on beaters, beater speed, fan speed and the feed. A higher beater speed shall give more neps, if the material is not moved out of the beating area effectively. So check the setting of leather flaps, stripper knife, etc.

Excessive of soft wastes fed and cottons with more immature fibres also contribute significantly for neps in opened material. It is therefore necessary to put a limit on the soft wastes added in the mixing. Selection of cottons considering the maturity coefficient and the micronaire values are very important where the presence of neps is critical for the end use.

Longer length of conveyor pipes and bends in pipes are also responsible for generation of neps. Further the smoothness of surface inside the pipes is an essential factor for getting nep-free yarns. The inner surface of pipelines gets rough because of wax coating and trash getting embedded with the wax. Therefore, periodic cleaning of the inner surface of the pipelines using rough cloth like Hessian is essential.

5.1.3   High variability in the delivered hank

The variations in delivered material are directly linked to the variability in feed and improper synchronization of the machines in blow room line. Improper levels in the hoppers, improper action of feed regulators, viz., cone drums, pedals, photocells, direct driving gear motors, etc., need to be checked and rectified.

5.1.4   Formation of cat's tail

If material movement is less and cottons are over beaten, we get this defect. By controlling the feed, sharpening beater edges, increasing fan speeds, increasing the air inlet below the grid bar area of cotton entry, closing the striping knife and beater setting shall avoid cat's tail. The very important step in avoiding cat's tail is to avoid chocking of materials in beaters.

Wet material if fed to blow room results in cat's tail. It may be due to excessive use of cotton-spray oil or water, not allowing for complete conditioning, dropping of water particles from Bahnson fans, leakage from roofs or dropping of steam condensates in cold season. It is essential to check the condition of mixing before feeding to blow room and prevent any water leakage in the department.

5.1.5   Conical lap

Conical laps are due to either higher quantity of cottons coming on one side of the lap or due to unequal calender and rack pressures in scutchers. Ensure equal opening of air inlets under grid bars, replace torn leather lining at the cage, clean the cage thoroughly with emery paper, make the pressure on lap spindle uniform on both the sides, remove the pedals and clean thoroughly and check the pedals where it rests on fulcrum and also pedal fulcrum bar.

5.1.6   Lap licking

Lap licking can be due to excessive addition of soft wastes in mixing, higher rack pressures, lower compacting of laps and excessive dampness in cotton. In case of polyesters, this problem shall be mainly due to static charges and higher bulk of fibres. The problem of lap licking can be reduced by increasing the pressure on calender rollers, reducing the pressure on racks, increasing the quantity of antistatic, use of roving ends or lap fingers behind the calender roller nip, blocking of top cage and by reducing the length of lap.

5.1.7   Patchy lap

Patchy lap is a result of unopened tufts. Ensure that the mixing is opened thoroughly and increase opening points if feasible. Check tuft size at the delivery of each beater and adjust the setting between feed roller and beaters, reduce the gauge between evener roller and inclined lattice, clean the cages and increase effective suction at cages.

5.1.8   Holes in lap

Damaged cage surface and blocking of the perforations in the cage are the normal reasons for a hole in the lap. Check the cages for damage and reduce tension draft.

5.1.9   Soft laps

In order to overcome the problem of soft laps increase the calender roller pressure. Very dry climate can also contribute for the soft laps. Check the humidity and maintain around 55% R.H.

5.1.10 Ragged lap selvedge

Check for the rough spots on the sides of the feed plates, leather linings for the cages and keep the edges of the scutcher clean.

5.2   Carding

5.2.1   Patchy web and holes resulting in uneven card sliver

This problem may be due to loading on the cylinder, damaged/pressed wire points in cylinder, doffer or flats, waste accumulation below cylinder under casing or defective under casing, heavy feed to cards, unopened cotton lumps from the cylinder-flat region, damp material and too fine or immature cottons. We need to analyse and find the exact reason prevailing at the time of study and take appropriate corrective action.

5.2.2   Cloudy web

The main reasons for cloudy web are improper setting of feed plate to licker-in, missing or damaged teeth in licker-in, damaged wire points on cylinder or doffer, doffer not set parallel to cylinder, cylinder and doffer not ground accurately, improper settings of cylinder to flat, improper settings of cylinder to back plate, insufficient feed roller grip, excessive fly under doffer and inadequate grinding frequency.

5.2.3   Singles

Singles in the sliver may be due to lap licking, less feed in chutes, part of carded web getting sucked by the waste extractor, damaged doffer wire and direct air currents hitting the web.

5.2.4   Sagging web

Sagging web in a card may be due to insufficient tension draft, very high humidity, worn out key in the calender roller gears, blunt doffer wire, dirty trumpets, heavy material fed to card and inadequate calender roller pressure.

5.2.5   Bars in card web

The bars in card web are mainly due to damaged cylinder wire, doffer wire or back plates across width, improper setting of back plate, eccentric licker-in surface, eccentric doffer surface, eccentric movement of stripping roller and redirecting rollers.

5.2.6   Irregular selvedge in card web

The reasons for irregular selvedge are too wide settings of lap guides or feed guides, damaged clothing at the side, air escaping at the sides of cylinder and excessive accumulation of lint at the sides of the under-casing.

5.2.7   High card waste

Damaged under casings, higher flat speed, wider front plate setting, closer setting of flats, higher pressure in suction unit and fibres getting ruptured are the main reasons for higher card wastes.

5.2.8   Low nep-removal efficiency

Blunt wire points, too wide setting between feed plate and licker-in, uneven settings, burrs in front plate/back plate and card wires of coarse type contribute for low nep removal efficiency.

5.2.9   Higher U% of sliver

Worn out parts, eccentric or wobbling gears, gears meshing too deep, oblong bores of gears, loose keys and key ways, insufficient tightening of gears, bad trumpets, waste accumulation in material patch, improper settings and loading of fibres on cylinder and flats contribute for higher U% of the card sliver.

5.2.10 Medium and long-term irregularity

Medium and long-term irregularity are due to insufficient feed roller grip, higher yard to yard (metre to metre) CV% of lap, the height of material in the chutes not controlled properly, worn out gears in feeding zone, obstruction in the movement of aprons during doffing, higher tension draft, damaged back plate and improper working of autoleveller.

5.2.11 Bulky sliver

Trumpet of a very large size and lower calender pressure are the main reasons for a bulky sliver. When the bulk of fibre is more (higher denier or higher micronaire) the sliver will be bulky.

5.2.12 High breaks

Very small trumpet, worn out trumpet, uneven sliver with bunches of fibres, worn out gears, damaged clothing, air currents disturbing the web, improper temperature and humidity and a very high tension-draft are the normal reason for high sliver breaks.

5.2.13 Fibre rupture at cards

The fibre rupture takes place at cards due to improper opening of tufts at blow room, heavy feed, closer setting between feed plate and licker-in, very dry cotton with moisture content of less than 4% and very high cylinder speeds.

5.3   Combing

5.3.1   Inadequate removal of short fibres and neps

Check head to head and comber to comber noil percent variations and check the individual heads for web defects, such as uncombed portions due to slippage under feed roller, slippage of fibres under detaching rollers, plucking of fibres by half lap from nipper grip, web disturbance due to air currents because of defects in brush or/in aspirator. Check the machines thoroughly for bent and hooked needles on half lap and top comb, broken needles, nipper grip, feed roller grip, condition of detaching roller cots, condition of the gears driving bottom detaching rollers and damaged air seals in the aspirator box.

5.3.2   Short-term unevenness

Prominent piecing waves, drafting waves, uneven fibre control due to worn out top roller cots in draw box, eccentric rollers in drafting/detaching field, play in draw box drive, high or low tension draft, and improper settings are the main reasons for short-term variations in a combed sliver. Check U% and make use of spectrogram diagram to identify the source of the problem.

5.3.3   Piecing waves

Piecing waves posses a periodicity with the wave length corresponding to one combing cycle. The normal reasons are incorrect detaching roller timing, play in drive between two detaching rollers, web defects repeating in every combing cycle, higher tension drafts between detaching rollers and table calender roller, improper setting of sliver guides on table and play in drive between draw box and coiler.

5.3.4   Drafting waves

Drafting waves are caused by too wide settings in draw box, worn out top roller cots and end bushes, eccentric bottom rollers and bearings and ineffective loading.

5.3.5   Hank variations

Single, double or uneven working of sliver on table due to improper selection of tension draft, rough surface of the sliver table, variation in the feeding lap, lap licking while lap unwinding, etc., are main reasons for variation in hank. If in between comber variations are high, check the combers for variations in lap roller feed per nip; draft wheels on draw box, tension drafts at tables, draw box and coiler and noil level variations.

5.3.6   Higher sliver breaks at coiler

The prominent reasons for higher breakages at coiler are sliver guides with rough surfaces, coiler calender rollers having eccentricity or jerky motion, high tension draft, improper balancing of sliver stop motion working on gravity principle, worn out gears, worn out key and key ways, excess parallelization of fibres in the sliver, improper condensation are the main reasons for sliver breaks. Check whether tension draft between draw box calender roller and coiler calender roller is too high causing stretching of sliver, or too low causing slackening of sliver. Check balancing of sliver break stop-motion and ensure that it presses against the sliver very lightly.

5.3.7   Frequent coiler tube choke-ups

If the coiler tubes are loaded with wax/trash, the sliver gets chocked. Clean the coiler tube with a rough rope. If cans are over filled, or the can spring is forcing the sliver to coiler plate, the choke up shall take place.

5.3.8   Web breakages at draw box

Burrs or accumulation of wax/trash particles at trumpet, too much spreading of web, defects in gear wheels, improper tension drafts are the main reasons for breakages in the draw box zone. In cases where the top rollers are buffed badly, the cottons shall stick to top rollers and lap resulting in breakages.

5.3.9   Breakages at sliver table

Waxy and rough surfaces of the table, improper tension drafts and piecing waves are the main reasons for breakages on the sliver table.

5.3.10   Breakages on comber heads

A tight or slack web, improper positioning of web trays, unclean web trays, burrs in calender trumpets, improper calender trumpet (heavy or light), improper functioning of clearer rollers in detaching section, piecing waves and the trumpets set too away from the nip of calender rollers are the main reasons for breaks at comber heads.

5.3.11   Detaching roller lapping

Rough or waxy surfaces on top roller cots, improper functioning of clearer rollers, too wide a setting of web guides are some of the reasons for lapping on detaching rollers. As the detaching top rollers tend to bend at the centre because of loading at both the ends, taper buffing is recommended.

5.3.12 Excessive lap licking and splitting

Improper tension drafts and roller setting, excessive draft in the lap former, uneven lap and tight winding while lap preparation are the main reasons for lap licking and splitting.

Note: Combers are very sensitive to changes in temperature and humidity, and hence it is essential to maintain the required temperature and humidity. In the majority of cases the bad working is attributed to fluctuations in temperature and humidity.

5.4   Draw frames

5.4.1 Improper sliver hank

Check the hank of input slivers and ensure they are as per plan. Check the draft wheels and ensure that the wheels are put to get the required draft. Check the functioning of autolevellers by sliver test method, (i.e., A%) and ensure that the input voltage is as per norms. Check the pressure on top rollers and ensure them to be as per norms.

5.4.2 Uneven sliver

Check the condition of top and bottom rollers, setting of rollers, the pressure on the top rollers, the condition of the end bushes of the top rollers, worn out or loose wheels, ensure even feed material. Make use of Uster spectrogram for identifying the source of the problem. Ensure that the slivers do not hit the can surface while getting filled. A bad quality spring in the can make the sliver tilt and spoil the same.

5.4.3 Singles

Stop motion failures or delay in action of a stop motion are the main reason for singles in draw frame as a can run out is not noticed by the tenter. A very high suction power of pneumafil sucks good fibres and can result in singles. The singles for a short length can also be due to partial lapping on rollers.

5.4.4 Cuts in sliver

Cuts in sliver are mainly due the settings not matching to the fibre staple length. However cuts can also come due to eccentric rollers, slips in the middle top roller, defective top roller weighting, worn out end bushes, worn out keys and studs in draft gear assembly and eccentric coiler shaft drive and grooved calender rollers. The cuts in slivers can be best observed by twisting around 2 m of sliver by hand (Fig. 5.1).

5.1 By twisting around 2 m of sliver by hand, one can identify if cuts in sliver are there.

5.4.5   Good fibres in suction waste

Too close a setting of suction nozzle and a very powerful suction are the reasons for good fibres going in suction waste.

5.4.6   Improper coiling

Non-centering of can and eccentric bottom plate is the main reasons for improper coiling. Also the speed of the can and coilers are to be synchronized to have the required spacing of coils in the can.

5.4.7   Higher breakages

Check the hank and uniformity of sliver, the sliver condensation, condition of gears, the tension draft and ensure smooth surface which comes in contact with sliver/web. Ensure the temperature and humidity to be as per requirement. Check for the surface of cots; if it is rough, it is likely to lap. Similarly, if the top roller pressure is very high, there shall be lapping on top rollers.

5.5   Speed frames

5.5.1   Higher U% of rove

Inadequate top arm pressures, improper settings, worn out gears/bearings, grooved top rollers, tilted top rollers, wrong selection of condensers, worn out aprons, poor cleaning of draft zone, higher stretch, uneven feed material, sliver splitting in creel, jerks in creel movement, vibrations in the machines are some of the reasons for higher U%. It is always essential to refer the spectrogram and check the spindle and the feed material before taking any action.

5.5.2   Higher breakages

Uneven material, worn out parts, vibrations, insufficient twist, improper draft distribution, fluctuations in temperature and humidity, rough surface in the flyer tube, improper build of bobbins, improper piecing of draw frame sliver, uncontrolled air current, etc., are the main reasons for breakages at speed frames.

5.5.3   Soft bobbins

Soft bobbins are due to finer hank, may be due to singles or a finer drawing hank, less number of turns per pressure, and the shift on cone drum (building mechanism) faster than required, a lower TPI and a lower relative humidity.

5.5.4   Lashing in

When a broken end joins to an adjacent end, we get lashing-in (see Fig. 5.2). Fixing of separators, setting the suction tubes near to the front roller nip shall solve this problem. Moreover we should work towards zero breaks.

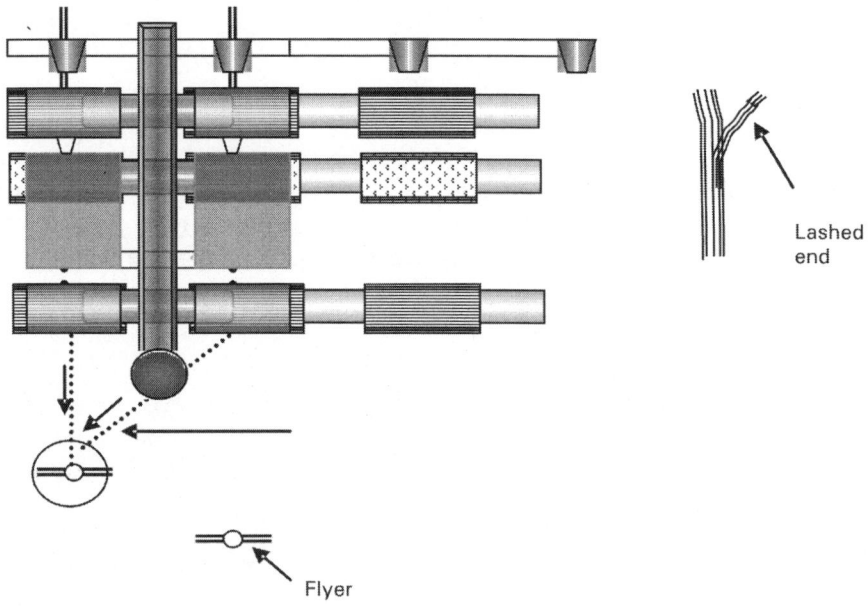

Lashed end

Flyer

5.2 Lashing in.

5.5.5   Hard Bobbins

This is due to a coarser hank; may due to doubles, coarser draw frame hank and lower pressure in top arms. Hard bobbins are also due to higher twist, lesser movement of belt on cone drums, higher turns on the flyer presser, shifted cots in the back zone leading to low pressure and higher relative humidity.

5.5.6 Oozed out bobbins

The malfunctioning of reversing bevels in the builder motion, stopping the machine when the bobbin rails are in extreme positions and jumping bobbins are the normal reasons for oozing out bobbins (Fig. 5.3).

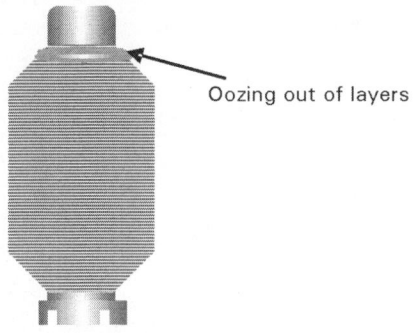

Oozing out of layers

5.3 Oozed out bobbins.

5.6. **Ring frames**

5.6.1 Ring cut cops

Ring cuts are due to the cop diameter becoming more than the limits prescribed by the ring diameter. The possible reasons are the count being very coarse, the build not adjusted properly, a bigger ratchet wheel, less number of teeth on ratchet wheel being pushed each time, loose ratchet wheel, vibrating spindles, non-alignment of rings in the centre of spindle axis, use of a lighter ring traveller, jammed poker bars, insufficient pressure on top rollers resulting in coarser count, etc. (see Fig. 5.4).

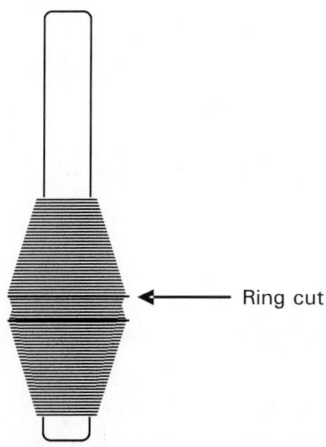

Ring cut

5.4 Ring cut cops.

5.6.2   Hard twisted yarn

Hard twisted yarns are due to count becoming very coarse because of inter doubles or lashing in, front roller delivery becoming less due to a loose roller, worn out threads in fluted roller joints, failure of delay drafting mechanism in G5/1 ring frames, loose timer belt driving the front roller in G5/1 ring frames, traverse going out of drafting area, etc.

5.6.3   Uneven yarn

Uneven yarn are due to uneven feed material, improper settings in drafting zone, worn out cots, worn out aprons, improper selection of spacer, low pressure in top arms, eccentric fluted rollers and cots, jammed arbours, vibrating spindles, jammed bobbin holders, improper distribution of drafts between break draft zone and main zone, improper cleaning of drafting zone, lapping in adjacent spindles, traverse going out of drafting area, etc.

5.6.4   Soft twisted yarn

Soft twisted yarns are due to loose tapes, worn out tapes, jammed spindle bolsters, loose bobbin on spindle, jammed jockey pullies, spindle button missing, etc.

5.6.5   Higher hairiness

Higher hairiness are due to worn out rings, worn out travellers, worn out lappet hooks, worn out separators, higher spindle speed, improper selection of traveller, variations in fibre lengths, lower humidity in the working area, too big a balloon, vibrating spindles, etc.

5.6.6   Lean built cop

Lean built cops are due to very fine count, excessive breakages on a particular spindle, very high chase length, smaller ratchet wheel, etc. Normally the eccentric spindles and non-centering of lappet hooks are the one who contribute for high breaks and then to lean built cops.

5.6.7   De-shaped cops

Not attending to breakages in time, non-attending to creel runs out in time, excessive breakages are the main reasons for de-shaped cops (see Fig. 5.5). The cops shall not have the normal build and the yarn content shall be less. The winders refuse to run such bobbins as it reduces their efficiency.

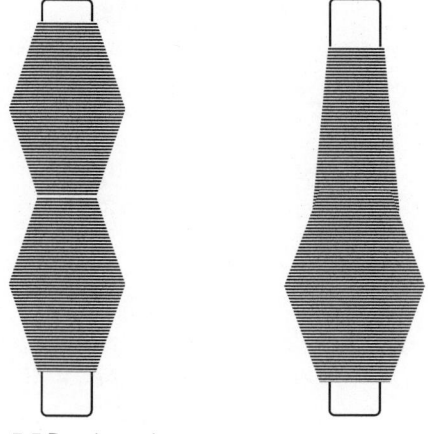

5.5 De-shaped cops.

5.6.8   Sloughing off at winding

Loose-built soft bobbins, very low chase length, improper combination of winding and binding coils are the normal reasons for sloughing off at winding.

5.6.9   Undrafted end

High twist in rove, lower break draft, low top arm pressure, higher humidity, smaller spacer, channeled aprons and cots are the normal reasons for undrafted ends in spinning.

5.6.10 Higher thin places

Excessive draft, setting wider than required, worn out gear wheels in drafting zone, jerks in working, eccentric movement of cots and fluted rollers, partial lapping on drafting rollers, higher stretch between bobbin holder and drafting zone, broken roving guides are the normal reasons for higher thin places.

5.6.11 Idle spindles

The idle spindles add cost to the manufacturing without producing. Some mechanical reason prevents the spindle from being utilised by the sider. The main reasons are non-creeling of bobbins in time, tape and apron breakages, oil flowing to rollers from gears and broken/missing spare parts.

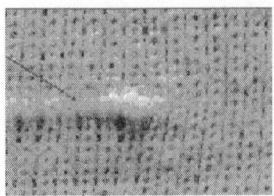

5.6 Slub damaging the fabric.

5.6.12 Slub

Slub is a thick, uneven twisted place in yarn. It can happen due to improper mixing of fibres, fibres with too much variations in fibre lengths, improper opening, low pressure in drafting roller, inadequate drafts applied, lower setting of drafting rollers for the fibre length in use, damages in card wire points resulting in bunches of fibres, lashing in or lapping in spinning, damages in the draft gear wheels, slippage of rollers while drafting (Fig. 5.6).

5.6.13 Crackers

Cracker is a short undrafted portion in a yarn that gets straight when pulled. If a ring frame working on a coarse count for a pretty long time is taken for fine counts, we get this type of problem. This is because of fine channeling on the top cot because of the thick rove working. In order to avoid this problem, we need to buff the cots lightly before changing from a coarse count to fine count.

5.7   Winding

5.7.1   Improper splicing

Improper splicing leads to opening up of the splices in working leading to breakages and loss of efficiency. The reasons for improper splicing are low air pressure for splicing, improper mingling chamber and improper setting for opening and splicing. The prism selected should match the count and the direction of twist along with the fibre length. Normally spinning mills run different counts on the same winding machine, but shall not be able to match the prism size, the air pressures, timing, etc., required for the yarn every time. The splice opening problem is more in coarse counts.

5.7.2   Electronic yarn clearer (EYC) failures

Fluff accumulation in the measuring slot, low input voltage, blunt cutters, jammed cutters, loose fitting of PCB are the main reasons for EYC failures or malfunctioning in a winding machine.

5.7.3   Double end

Failure of EKP (Electronic Knotter Programmer) in cutting the end after a splice/knot, failure of bobbin conveyor belt are the main reasons for double ends found on cones.

5.7.4   Stitches

Variation in cone holder settings, vibrations in cone holder or drum, loose fitting of cones and damaged drums are the normal reasons for stitches. The static electricity generated during winding, especially creates stitches in manmade fibre yarn winding, which are prone for static charges.

5.7.5   Soft and bulged cones

Very low tension, yarn going out of the tension disc, fluff accumulation between tension discs, improper rotation of tension discs are the normal reasons for soft or bulged cones.

5.7.6   Sunken nose/base

Improper fitting of paper cones on cone holder, improper size of paper cones and improper setting of cone holder are the main reasons for the sunken nose or base.

5.7.7   Weight variation between cones

Variations in tensions between drums, improper setting of conometer or diameter on cones, malfunctioning of drum sensor, too much variation in yarn count are the normal reasons for weight variation between cones.

5.7.8   Shade variation within cones

A mix up of yarn from different mixing, variation in day to day mixing preparation and addition of soft wastes, ring cuts and abrasion polishing a part of yarn, very slack tape resulting in a very low TPI, contamination of oil or grease while spinning, exposing the ring cops to smoke or fumes are the normal reasons for shade variation with in cones.

5.8   Rotor spinning

5.8.1   Neppy and uneven yarn

Dust accumulation within rotors, mark in rotor grove, damages in rotor covers, lapping on opener roller, damaged wires on opener roller are the main reasons for neppy and uneven yarn in rotor spinning.

5.8.2   Stitches

Stitches on cheeses are mainly due to lapping on the cradle sides, choke up in the traverse path, lapping on drum, damaged cradle bearing, damaged bobbin holder and snap in traverse bar belt.

5.9   General

5.9.1   Stains

The stains are due to various reasons. The oil stains are mainly due to oil leaking and falling on the cottons or floor. This can happen due to excess oiling, not wiping out the excess oil and not properly cleaning the parts after oiling, oil spilling out due to chokes up in oil path not cleaned properly, worn out oil seals, improper covering, allowing cottons to fall on floor, improper selection of lubricating oils and grease, etc.

The workers need be trained to clean the oil path before oiling any part (Fig. 5.7).

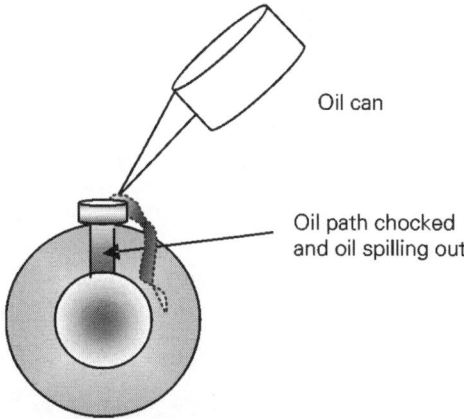

Oil can

Oil path chocked and oil spilling out

5.7 Chocking of oil path.

The oils need to be tested before accepting as to whether it is washable or not. There are some oils which make a permanent stain on cotton that is not washable, but there are some oils that are washable with the use of normal soap or detergents.

The stains also can be due to sweat or dirty hands of the operators. Normally these are washable unless otherwise the hands are oily. The rusted steam or water pipes if any in the production area also can cause permanent stains on the cottons.

Unusual blackening of yarns is seen in case of any v-belt burning off or fire accidents. The black smoke deposits on the yarn. It is therefore suggested to keep all the yarn separately in case of any fire accident in the shed and deliver it for further process after verifying the shade.

5.9.2   Contaminations

The reason for contamination is poor work practices. This is discussed in detail in Chapter 6.

6

Normal complaints from customers

While spinning yarns, the mills might do some mistakes either because of miscommunication, improper knowledge of the systems, ignorance, negligence or by over confidence. Even in the best managed companies, there shall be some poor quality and the customer observes this. In a manufacturing unit, normally some sample is tested and the lot is passed if the sample is okay, or sometimes it is in boarder. But the customer checks everything 100% and through out its life, as he/she uses the material number of times. The customer is, therefore, very clear on the level of quality received irrespective of the claims made by the producer about their quality level. The defects or the poor quality might escape the inspectors while checking the quality, but cannot escape the ultimate customer.

The customers might register a complaint or might not. It is left to the wishes of the customer, whether to register a complaint or not. But the customer shall show the problem to some of his/her close people and express his/her dissatisfaction. Customer has two options; registering a complaint and putting a claim or changing over to a new supplier. The customers normally do not wish to make a complaint. They would be searching for a source from where they can get fault free materials that does not affect their process or image.

The customers, when they are not happy, but have a confidence on the supplier shall make complaints. If the customer is not having confidence on the supplier, shall change over to another, as complaining and fighting for compensations is costlier compared to changing over to a new supplier. It is observed that only 10–15% of the customers make complaint whereas the remaining prefer to change the customer without making any noise. The customers those complain are interested in the mill to improve and cater them with better products. They are normally happy with the dealings and the quality, but shall caution the spinner when the quality is going out of controls so that suitable preventive actions could be taken. These few customers normally contribute for the major portion of sales.

The customers complain or express their unhappiness not only on the quality of the products but also on the services that include the way in which the customers are treated, their queries and calls are responded, information provided, the packing and forwarding, etc. In this chapter we are discussing on the product related complaints specific to yarns. Some of the normal complaints a spinner receives are the shift in count, count variation, low strength, high or low twist and variations, unevenness, presence of trash and kitties, hairiness, barre, objectionable faults, mix ups, stains, contaminations, winding quality and packing quality. Let us discuss some of the common complaints a spinner receives.

6.1   Count and count variation

6.1.1   Why customers are bothered about high count CV%?

The first parameter normally tested in any yarn is the count and the count CV%, followed by strength, twist and evenness. The count CV% is a very important parameter. We need to keep the CV% as minimum as possible as any variation means uncertainty. Very high fluctuation in yarn quality is an evil for any end use. Vijayakumar[6] opines it as better to keep same level of yarn quality (say around 25% Uster standards) by strict quality control than achieving 5% Uster standard, but without consistency. He observes that consistent quality is very much appreciated by the clients.

Count variation means variation in linear density of yarn leading to variation in diameter. Higher count CV% indicates higher variation in count between bobbins. The variations in yarn diameters affect the appearance, dye pickup and also feel of fabric. Depending on the end use of the fabric the effects have different levels of criticality. Following are some examples:

- Streakiness in woven as well as in knitted fabrics
- Barre in circular knitted fabrics
- Corkscrew effect when doubled
- Uneven sizes in case of socks
- Length variations in cone when doffed with a fixed weight resulting in more hard wastes in warping

Let us discuss the impact of count variation on the end products.

(a) Yarns for knitting T-shirts

- A deviation in average counts results in higher or lower weight of fabric. If the count is finer the weight will be lower and is not accepted especially in T-shirts as the comfort depends on the fabric density. If the count is coarse, we need more yarn to knit the garments, and hence, it is a loss to the customer. The count also affects fabric feel and wear-comforts.
- In case where the count is coarser than the gauge of the knitting machine, it leads to problems in working also.
- Variations within and between cones result in streaky or barre effects.

These effects are not accepted by the customer as it gives an ugly appearance.

(b) Yarns for socks knitting

- In socks knitting, we normally use less number of cones like one cone or two in a creel. As the socks are very small, normally a sock produced belong to either one bobbin or two bobbins. Hence count variations in yarns will not give any problem of barre, as we get in T-shirts, but the weight of socks and the dimensions of socks get changed.

(c) Yarns for sweater knitting

- Normally, sweater knitters take double or multifold yarns, and hence chances of getting higher count CV% are less.
- If the average count varies, the fabric weight shall vary significantly.
- The variations in fabric density change the entire look of the sweater.

(d) Yarns for underwear knitting

- A deviation in average count results in variation in weight of fabric.
- Deviation in count affects the wear-comfort and fabric feel.
- Variations within and between cones give rise to barre effect, however, as the fabrics are used for underwear, this is not considered as critical.

(e) Yarns for knitting liner materials

- Variations in average count results in higher or lower GSM of the fabric. This gives problem in coating as it becomes uneven.
- Within cone and between cone variations in count gives a barre appearance and results in cracks after coating.

(f) Yarns for weaving apparels

- A deviation in average count results in higher or lower weight of fabric, and also affects the fabric feel and wear-comforts.
- A coarser count results in lesser fabric length for the same weight of yarn, leading to a loss to the manufacturer.
- Variation within cones and between cones results in streakiness in warps or weft-bands in weft. The width of the bar depends on the yarn content in the cop.

(g) Yarns for heavy fabrics and industrial applications

- As normally doubled and multifold yarns are used for heavy fabrics, there shall not be many problems due to count variations.
- Any shift in average count results in a shift in fabric weight, which shall create problems in further processes and in the strength of the fabric.
- A variation in count can lead to lower breaking strength because of thin portions which is critical for industrial applications.

(h) Yarns for towels

- A deviation in average count results in variation of towel weights. The variation between cones and within cones does not create problems like plain cloths for apparels as the designs and the piles cover up these variations.

(i) Yarns for carpets and furnishings

- The shift in average count is more critical as it affects the weight of the fabric.
- The variations within cones and between cones are not considered as critical.

(j) Yarns for sewing threads

- This is not a critical issue as the sewing threads are used as a single thread, and not in group as in knitting or in weaving.
- There are chances of getting cork screw effect in the yarn.

6.1.2   Major reasons for count variation

The higher count CV% in a yarn can be attributed to:

- Machinery conditions
- Work practices
- Working parameters

The preparatory machines from draw frames onwards directly impact the count variations, whereas the earlier machines, although produce uneven materials, do not directly make an impact in the yarn. The variations can be corrected at draw frames by suitable doublings and installing auto-levellers, but the variations caused in and after draw frames cannot be corrected. The draw frame plays a major role in controlling the count variation. The studies done by V. Ramachandran shows that 50% of the contribution for count variations is from draw frames, whereas blow room and cards contribute for 13%, roving frame for 20% and ring frames for 17%. Therefore, draw frames need proper monitoring and maintenance to reduce the complaints of count variations and barre.

Let us discuss the contribution of ring frames, speed frames and draw frames in getting higher count CV%.

Contribution of ring frames

Machine conditions – 1

- Low top arm pressures resulting in uneven draft. Uneven draft leads to uncontrolled counts of yarn. Low top arm pressures may be due to air leakages in pneumatically weighted drafting systems, worn out springs and saddles, the top rollers of lower diameter, improper alignment of top rollers, improper setting of top arm pressure, etc. This leads to uncontrolled count variation.
- Variation in pressures between top arms leading to uneven drafts.
- Top rollers not aligned properly resulting in slippage (see Fig. 6.1). The slippage gives lower pressure and uneven drafting leading to short-term count variations.

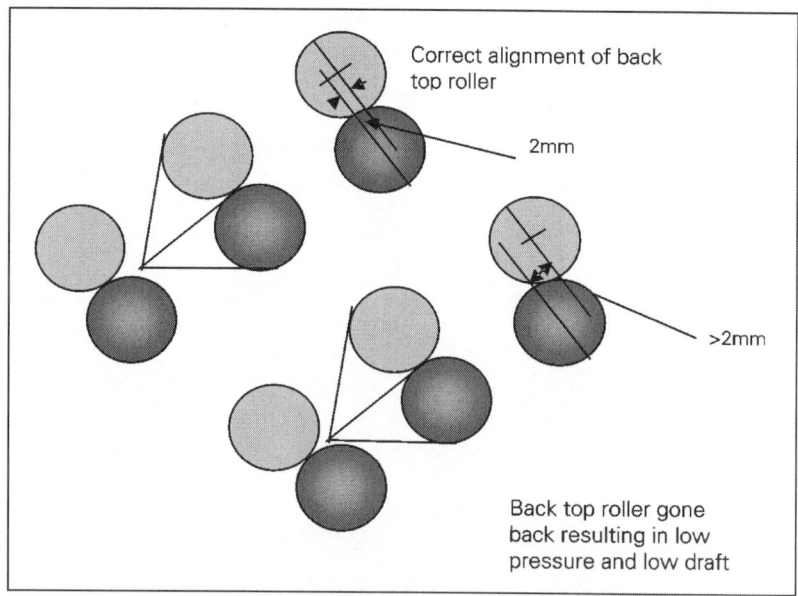

6.1 Alignment of top rollers.

- Bobbin holders not revolving freely introducing stretch in rove. The stretch introduced is uncontrolled; hence the count produced also would be uncontrolled.

- Creel not spacious to hold big bobbins resulting in touching of bobbins when all big bobbins are creeled (see Fig. 6.2). When the bobbins touch one another, they restrict the movement as they shall be running at opposite direction at the place of touching. This normally results in breakages. In sometimes, when the rove is stronger and due to jerks in the machine or in the creel, the bobbins revolve and rove is fed to spindles, but there shall be uncontrolled stretch leading to fine yarns.

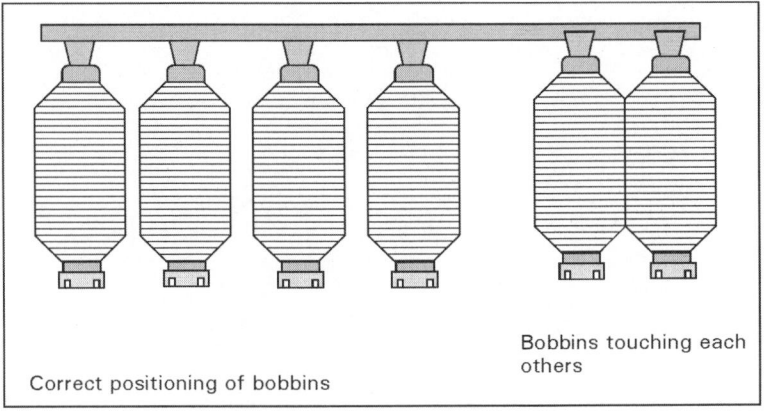

6.2 Bobbins in ring frame creel.

- Traverse not set properly allowing the rove to go to the extreme end of the cots. When the rove goes to the extreme end it might not get the grip of the rollers and there shall be lesser draft or no draft (see Fig. 6.3).
- Loose fitting of a gear in drafting zone (worn out pin) resulting in slippage. This problem is normally seen in the back bottom roller end wheel. Back roller, in spite of missing pin revolves because of the combined force of all roves that are being pulled by the front and second rollers. But the speed will not be as per our requirement.
- Worn out threads in back bottom roller joints. This leads to slippage of back bottom rollers resulting in lower and irregular drafts.

Ring frame work practices – 1

- Cross creeling of Bobbins leading to uneven stretch (see Fig. 6.4).
- Partial lapping on back bottom rollers not observed and cleaned (see Fig. 6.5): When a part of bottom roller is lapped, the rove running cannot get full pressure required for drafting. Hence it gives raise to coarse count. As the traverse is working, the pressure shall be getting changes as the rove moves near to the lapped portion and moves away from that place.

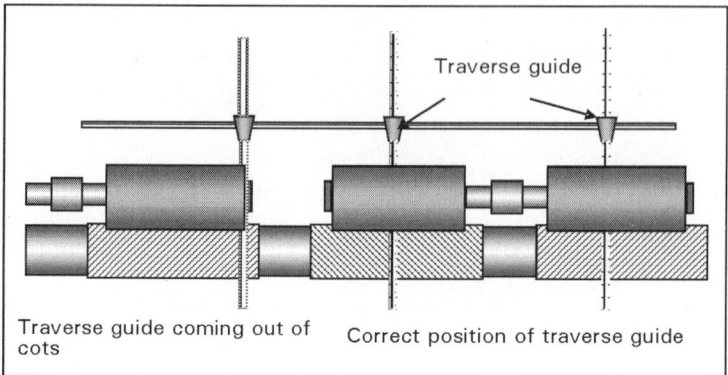

6.3 Positioning of traverse guide.

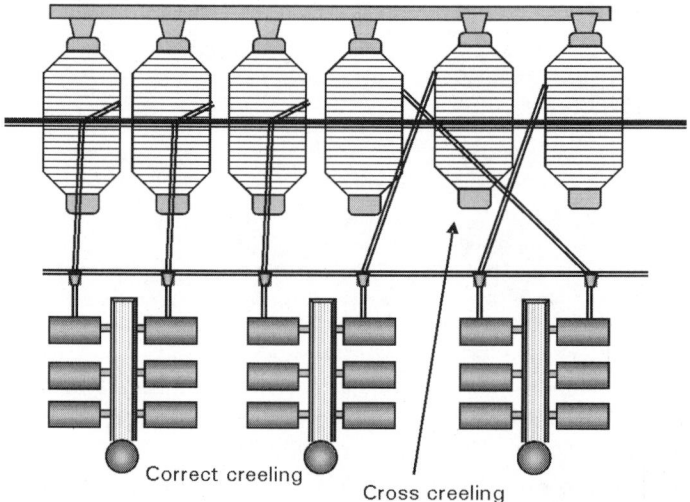

6.4 Creeling.

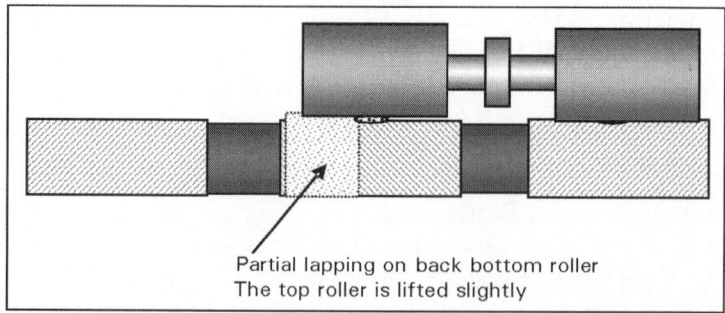

6.5 Partial lapping.

- Fluff accumulations in roving guide not removed: This can hold and remove the fibres from the rove and leads to stretching of the rove resulting in uncontrolled finer count.
- Keeping inter bobbins in stock for a very long time: The inter bobbins, if kept for a long time unused becomes soft. A soft rove gets stretched easily compared to other bobbins. These bobbins give raise to uncontrolled fine yarns that are irregular.
- Not following the channelisation strictly and creeling bobbins from different inters leads to confusions, and action cannot be taken when problem is detected. If the channelization is correct, the actions can be taken on the machines from where the problem was sourced. Technically if all the machines are set properly and are identical, there should not be any problem of count variation.
- Changing the pinions to correct the count by wrapping: It is a normal practice to check the counts periodically by taking samples from ring frames. In earlier days, about 3–4 decades back, there was a practice of correcting the pinion by seeing the wrapping on each side of the ring frames. Gradually the trend shifted to changing the pinion in group on all the machines at a time. The change in pinion shifts the average count by 2.5–3.3% depending on the number of teeth on the pinion. As this shift is very high, some people started changing the back roller wheel as it had more number of teeth compared to the change pinion, and the shift in average is restricted to 1% or less. Whatever may be the wheel we change, there shall be a shift in the average value. There is no answer for the questions as to what happens to the material already produced and that are in pipe line. Also we are not clear as to when the average shifted, as we only had corrected it last time by changing the pinion. It means the variations in hank or count seen is the natural variations. By changing the pinion we are adding one more variation.
- Overlapping of the rove while attending a running out bobbin can create doubles for 15–30 cm depending on the tail end left by the tenter while attending (see Fig. 6.6). This gets multiplied by the draft may be 20 or 35 times and we get coarse yarn of 3–10 m length.

Check points in ring frames for controlling count CV%

The following check points in ring frames can help in reducing the contribution of ring frames for count variations:

- Alignment of bobbins and the spindles and ensure no cross creeling.
- Whether the count of yarn is as per requirement and the variation is within limit or not? This check should be done immediately while starting a new count. The periodic checks shall help to detect any deviations or abnormalities.

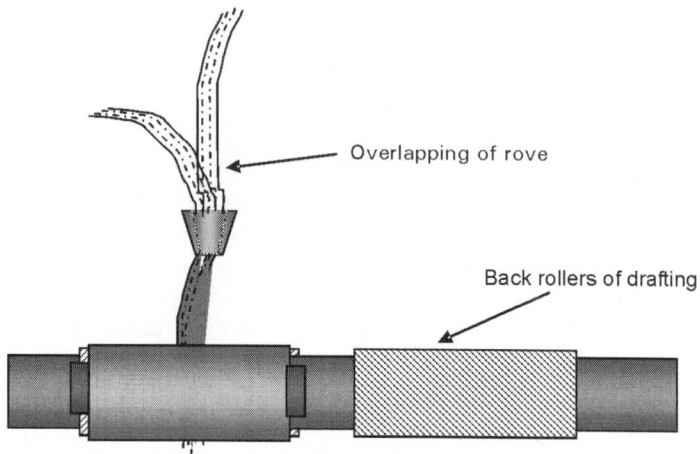

6.6 Overlapping of rove.

- Whether the condition of the machine parts that can contribute for count variations is good or not?
- Whether the breakages are in control or not?
- Whether the workmen are adequately trained or not?
- Whether the maintenance is done as per plans or not?
- Whether the colour codification and channelisation are followed properly as per plan or not?
- Whether the drafting zone and yarn path are always kept clean or not?
- Whether the temperature and humidity are maintained as per need or not?

Speed frame

A variation of 1 m in rove can lead to a variation of 20–30 m in ring frame leading to a count variation. This is depending on the draft adapted. Higher the draft more is the length of variation. The short-term variations or higher U% in speed frames lead to long faults like F, G, I and H faults and not count variation.

Contribution of speed frame machine conditions – 1

- Low or uneven pressure in the top arms lead to improper drafting and uneven rove, which gets enlarged into count variations.
- Misalignment of top rollers leading to slippage. This gives improper pressure and improper drafting.
- Unequal stretch between front and back row of the spindles is another main reason for variations. As the spindles in the front row are farther compared to back row, the stretch shall be high. It can be compensated

by raising the height of the flyer by adapting suitable false twisters (see Fig. 6.7).

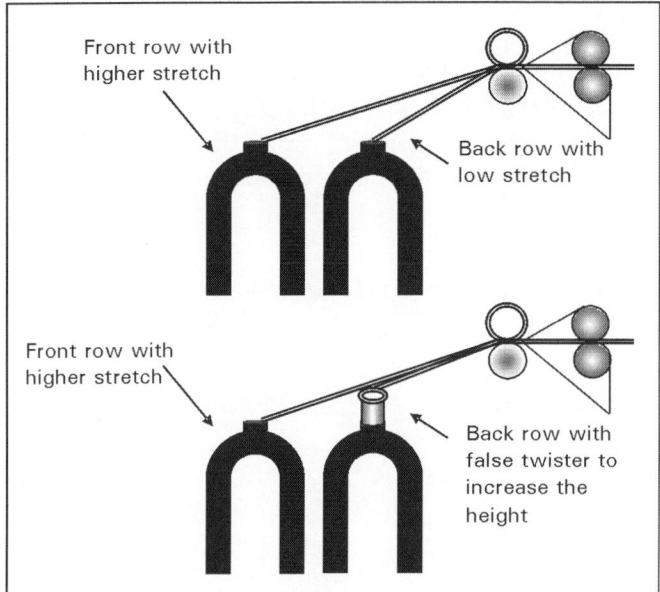

6.7 Stretch difference between front and back rows.

- Excessive stretch in creel in speed frames make the feed sliver fine and result in finer hank.
- Vibrating spindles and jumping bobbins give irregular stretch to the rove and lead to short-term variations. This is normally reflected as streakiness in the fabrics.
- Improper setting of builder motion leading to excessive stretch or loose ends. These not only lead to higher breakages in speed frames, and also count variations in spinning.
- Eccentric bottom rollers giving periodic variations in rove resulting in long thin and thick yarns of F, G, H and I fault in the yarn.
- Worn out gears and bearings especially in drafting area leads to cuts and thin places. This contributes for H fault in ring yarn.

Contribution of speed frame – work practices – 1

The following work practices are considered as bad as they contribute for variations:

- Cross creeling of cans
- Cutting cans and filling the blank spindles

- Overlapping the slivers while feeding a new can replacing a running out can
- Not attending the break in-time resulting in an undersized bobbin leading to loose ends
- Pressing the front top roller by hand to adjust a lose rove
- Not removing the singles and doubles while attending the breakages
- Not maintaining the separators leading to lashing in
- Checking the wrapping and adjusting the pinion in a running lot
- Not cleaning the flyers regularly allowing wax and trash accumulation that gives rise to uneven rove

Contribution of speed frame work parameters

- Low twist in rove leads to uneven stretch while unwinding at ring frames.
- Higher space between coils leads to uneven build resulting in stretch variation.
- A very low taper leads to oozing out of bobbins, and the oozed out layers give unequal stretch while unwinding at ring frame (see Fig. 6.8).

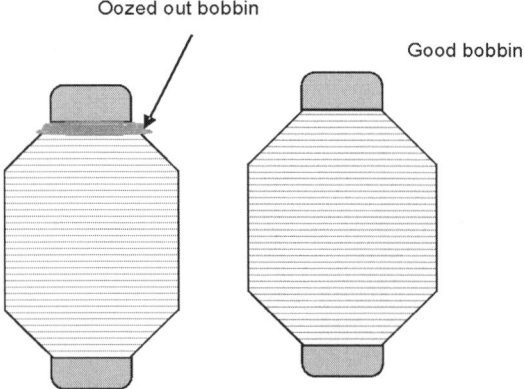

Oozed out bobbin

Good bobbin

6.8 Speed frame Bobbin.

Check points in speed frames – 1

The following check points and taking timely corrective actions should help in reducing the contribution of speed frames to the count variations:

- Whether the hank of rove is as per plan and the variations are with in norms or not?
- Whether the stretch is in control in all the spindles or not?
- Whether the breakages are within control or not?

- Whether the U% of the rove is as per the requirement or not?
- Whether all the parts of the machines are in good condition or not?
- Whether the settings and alignments of rollers in drafting zone are proper or not?
- Whether the drafting zone is always kept clean or not?
- Whether the top arm pressures are as required and uniform on all spindles or not?
- Whether the spacers, condensers, sliver guides, false twisters, etc., are as per plan or not?
- Whether the workers are following the work practices as specified or not?
- Whether the workmen are adequately trained or not?
- Whether the colour codification and channelisation are followed as per plan or not?
- Whether the maintenance is carried out as per plan or not?
- Whether the transportation of bobbins to ring frames is as per plans or not?
- Whether the temperature and humidity are maintained as required or not?
- Whether Bobbins are cleaned properly before putting on the machine or not?

Contribution of draw frames

As discussed earlier, the contribution of draw frames is the maximum to the problem of count variation compared to any other machine in the sequence of spinning. It contributes for more than 50% of the time. A variation of 1 m in sliver can lead to a variation of 200–300 m in ring frame leading to a count variation. Higher U% in draw frame can cause medium term variations in count at spinning.

Contribution of draw frames – machine conditions – 1

The machine conditions that contribute significantly for count variations are as follows:

- Improper setting of autolevellers
- Low pressure on drafting rollers
- Misalignment of top rollers
- Loose chain in the creel leading to uneven stretch of slivers
- Malfunctioning of stop-motions
- Excessive suction leading to sucking of good fibres from drafting zone (see Fig. 6.9)

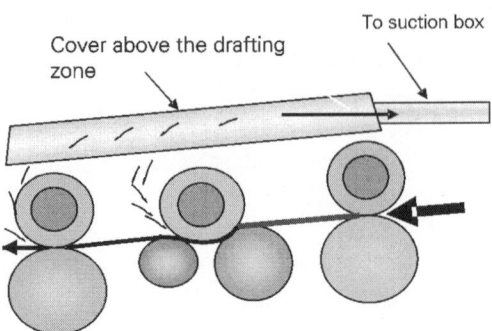

Cover above the drafting zone

To suction box

6.9 Loose fibres being sucked by suction fan.

Contribution of draw frames – work practices – 1

- Periodic checking the wrapping and adjusting the pinion in a running lot only adds to the count variation as the cans already produced are creeled in speed frames. They cannot be taken out, and we do not know when the shift in hank took place.
- Overlapping of slivers while feeding results in coarse count from 12.5% up to 16.67%. This depends on the number of doublings, either eight or six.
- Not removing the singles is a normal problem seen in all mills as taking out the cans, removing the singles and putting back is a difficult task, and also can lead to variations because of handling itself. It is therefore suggested to leave the sliver unpieced so that the speed frame tenter can remove those singles while attending creel breaks.
- Inactivating the stop motions are seen while attending some repairs or if short circuiting is found. They should be attended immediately and the production should be allowed to take place only after the stop motions are made active.
- Use of damaged cans and bent spring plates leading to disturbance of coils is a normal problem where the floor is not good and there is no suitable mechanism to transport the cans.

Check points in draw frames to control count CV% – 1

The following check points if followed and corrective actions taken can reduce the incidences of count variations to a great extent:

- Whether the conditions of the machine parts like drafting rollers, cots, end bushes, saddles, hosepipes, springs, belts, bearings, etc., are good or not?

- Whether the cans and springs are in good condition or not?
- Whether the wheels are put as per calculations or not?
- Whether the hank of sliver produced is as per requirement or not and the variations are with in limit or not?
- Whether all the stop motions are functioning properly or not?
- Whether the voltage variations are with in control or not for autoleveller draw frames?
- Whether the Sliver Test A% is with in norms or not for autolevelled material?
- Whether the colour codification and channelisation are followed as per plan or not?
- Whether the workmen employed are adequately trained or not?
- Whether the quality of cans and springs are as specified or not?
- Whether the cans are fully cleaned before putting in the machine or not?
- Whether the coiling is proper or not?
- Whether the trumpets are of correct size or not and whether they have smooth inner surface or not?
- Whether the surface of the sliver table (creel) is smooth or not?
- Whether the scanning rollers adapted are of correct size or not?
- Whether the temperature and humidity are maintained as per plan or not?

6.1.3   Attacking higher count CV%

After knowing the various reasons for count variations, we need to take suitable preventive measures to prevent the problem of count variations. Actions are needed at all places; and it is a continuous process. The introduction of autoleveller at cards and draw frames, better controls of humidity and temperatures, periodic training of the employees are essential factors contributing for preventing this problem. Avoiding frequent changes of counts and mixings also contribute greatly for overcoming this problem.

Draw frames and speed frame

As discussed earlier, the draw frames and speed frames combined together contribute for over 70% of the count variations. Therefore, we are discussing here on these two areas.

Draw frames

(a) Improper sliver hank:

(a) Check the hank of input slivers and ensure they are as per plan.

(b) Check the draft wheels and ensure that the wheels are put to get the required draft.

(c) Check the functioning of autolevellers by sliver test method, (i.e., A%) and ensure that the input voltage is as per norms.

(d) Check the pressure on top rollers and ensure them to be as per norms.

(b) Uneven sliver:

(a) Check the condition of top and bottom rollers, setting of rollers, the pressure on the top rollers, the condition of the end bushes of the top rollers and for worn out or loose wheels.

(b) Ensure even feed material.

(c) Make use of Uster spectrogram for identifying the source of the problem.

(d) Ensure the slivers do not hit the can surface while getting filled.

(e) A bad quality spring in the can tilts the sliver and spoils the same. Therefore, check the cans and spring quality periodically.

(c) Singles:

(a) Stop motion failures is the main reason for singles in draw frame as a can runout is not noticed by the tenter. Therefore, do not allow the machine to work if stop motions are not set properly.

(b) A very high suction power of Pneumafil sucks good fibres and can result in singles. Therefore, periodically check the suction fan wastes and adjust the suction.

(c) The singles for a short length can also be due to partial lapping on rollers. Check the surface of the rollers and maintain them clean. Periodically check the condition of clearer tubes and ensure that they are not hard and brittle.

(d) Cuts in sliver:

(a) Cuts in sliver are mainly due the settings not matching to the fibre staple length. Check the sliver quality immediately after setting the rollers or starting a new mixing. By holding about a metre of sliver in hand and twisting can help in identify the cuts faster. The use of spectrograms shall help in detecting the area from which the defect is coming.

(b) The cuts can also be due to eccentric rollers, worn out end bushes, eccentric coiler shaft drive and grooved calendar rollers. Therefore, proper maintenance of machine is very important.

(e) Good fibres in suction waste:

Too close a setting of suction nozzle and a very powerful suction are the reasons for good fibres going in suction waste. Periodically check the suction wastes and set the suction.

(f) Improper coiling:

Non-centring of can and eccentric bottom plate is the main reason for improper coiling. The supervisors should check the coiling after each maintenance activity. Also the speed of the can and coilers are to be synchronized to have the required spacing of coils in the can.

(g) Higher breakages:

(a) Check the hank and uniformity of sliver, the sliver condensation, condition of gears, the tension draft, and ensure smooth surface that comes in contact with sliver/web.
(b) Ensure the temperature and humidity to be as per requirement.
(c) Check for the surface of cots; if it is rough it is likely to lap.
(d) If the top roller pressure is very high, there shall be lapping on top rollers.

Speed frames

(a) Higher U% of rove

Inadequate top arm pressures, improper settings, worn out gears/bearings, grooved top rollers, tilted top rollers, wrong selection of condensers, worn out aprons, poor cleaning of draft zone, higher stretch, uneven feed material, sliver splitting in creel, jerks in creel movement, vibrations in the machines are some of the reasons for higher U%. It is always essential to refer the spectrogram and check the spindle and the feed material before taking any action.

(b) Higher breakages

Uneven material, worn out parts, vibrations, insufficient twist, improper draft distribution, fluctuations in temperature and humidity, rough surface in the flyer tube, improper build of bobbins, improper piecing of draw frame sliver, uncontrolled air current, etc., are the normal reasons for breakages at speed frames. Periodic checking of the machines and materials are very essential along with proper maintenance of temperature and humidity.

(c) Soft bobbins

Normal reasons for soft bobbins as discussed earlier are finer hank of rove; may be due to singles or a finer drawing hank, less number of turns on the flyer presser, the shifting of belt on cone drum (building mechanism) faster than required, lower twist than required considering the fibre fineness, length and the hank and a lower relative humidity. The supervisors should check the bobbins by pressing and ascertain as to whether it is hard or soft.

(d) Lashing in

Whenever an end is broken and joins to an adjacent end, we get lashing in. The lashing in shall normally be irregular as some of the fibres shall be thrown out into the air as there shall be no twist in the rove. The sider should remove the rove from the adjacent spindle and clean the lashed in portion. Fixing of separators, setting the suction tubes near to the front roller nip shall solve this problem. Moreover we should work towards zero breaks.

(e) Hard bobbins

This is normally due to a coarser hank of the rove. The coarse hank may due to doubles, coarser draw frame hank and lesser draft due to lower pressure in top arms. Hard bobbins are also due to higher twist, lesser movement of belt on cone drums, higher turns on the flyer presser, shifted cots in the back zone leading to low pressure and a higher relative humidity. The supervisor should feel the hardness of the bobbin and take action suitably. If the hardness could not be felt by hand, creel some bobbins on ring frame and observe for undrafted ends. Take report from the ring frame sider and decide on the action.

(f) Oozed out bobbins

The oozed bobbins are mainly due to malfunctioning of reversing bevels in the builder motion, stopping the machine when the bobbin rails are in extreme positions, and jumping bobbins. The oozed layers come very loose or might get stretched depending on their position and give raise to uneven counts. The empty bobbins are to be checked for perfect fitness before starting the machine and defective bobbins if any are to be discarded. The wearing out of notches to be checked periodically as worn out notches can lead to jumping of bobbins and oozing out. The workers need to be trained well for observing the position of bobbin rail before stopping a machine for any reason. Unauthorised or untrained persons should not be allowed to stop or restart the machines.

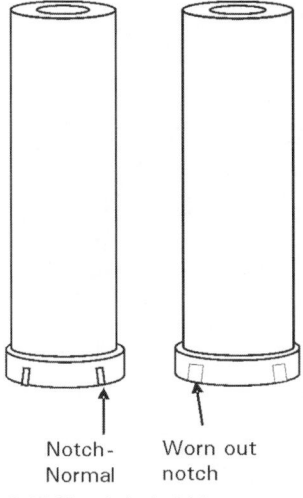

Notch-   Worn out
Normal   notch

6.10 Notch in bobbin.

6.2   Hairiness

High hairiness in the yarns is mainly due to presence of short fibres in excessive proportion, air currents disturbing the fibres in the yarn, rough surfaces abrading the yarn, generation of static charges while the yarn is being formed, etc. Improper selection of fibre denier or micronaire can lead to hairiness. Cottons with higher micronaire leads to more hairiness compared to the finer micronaire cotton for the same count and twist parameters.

Poorly centered spindles, spinning rings, anti balloon rings and yarn guides lead to inconsistent yarn tension. It makes the yarn touch or rub some metal parts, more than normal, resulting in hairiness. Rough surfaces due to damaged parts roughen the yarn; that might be the separators, the lappet eyes, rings and travelers, anti balloon rings, etc.

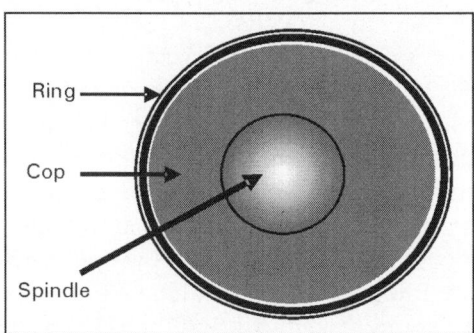

6.11 Poorly centered ring and spindle rubbing the yarn.

Open anti-balloon ring makes the yarn move out and come back again making an imbalance.

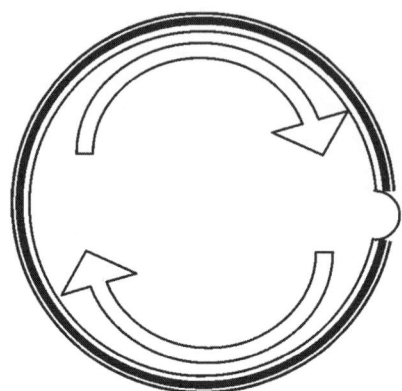

6.12 Yarn going out of Balloon control ring and again going in.

The clearance between ring and cop should not be too small. Traveller will cut the fibres protruding from the cop.

The fibres get electro-statically charged at high-speed spinning in the absence of proper humidification. The fibres tend to separate themselves from the yarn. Because the twist is holding some portion of the fibres in the yarn, a part of fibre stand out as hair.

If the traveller is very light, the twist propagation to the spinning triangle shall not be proper. The fibres do not get entangled inside the yarn and it leads to higher hairiness. The lighter traveller also results in a higher ballooning and rubs the yarn on balloon control rings and separators. The balloon going out of proportion makes the yarn rub with stationary parts of the machine and the fibres come out of the yarn. A larger balloon also faces higher resistance from the air which makes the fibres protrude out resulting in higher hairiness.

If the traveller weight is very high, the friction of the yarn shall be high and fibres come out of the yarn.

Yarn gets roughened in narrow yarn passage in the traveller. Any damage in the traveller thus leads to higher hairiness. The condition of traveller is therefore a very important factor for hairiness in the yarn. The studies show that the hairiness of the yarn gradually increases after certain time of traveller running, although we see that the traveller is in good condition. It was a normal practice in that mill to change the traveller after 6 days of working as 10% of the travelers were found burnt by that time. The controlled studies conducted by checking evenness of each cop doffed on selected spindles showed that the U%, imperfections and hairiness were at lower level for the first three days and started gradually

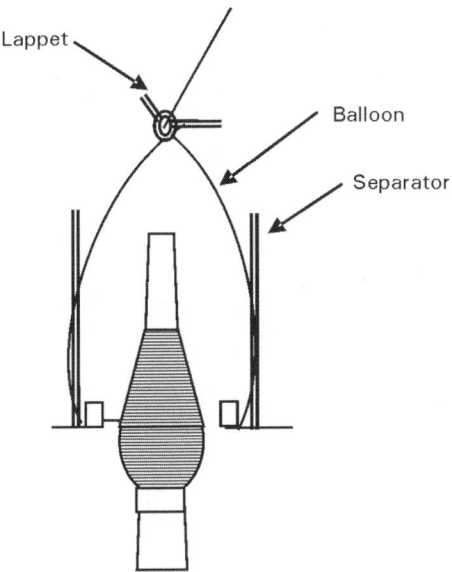

Lappet

Balloon

Separator

6.13 Excessive Balloon - Yarn crossing the separator and coming back.

increasing and became very high after fifth day. This indicated that burning of the traveller was the last stage in the traveller life, and we should not run them till they burn. After the studies, the mill started changing the traveller on every fourth day, and their quality level remained consistent.

The life of rings also plays a very important role in deciding hairiness. The rings get damaged because of the metal to metal friction between the ring and traveller. As the traveller is running and burning, carbon deposits on the inner surface of rings. It is necessary to remove that deposition by cleaning the rings periodically with clean dry cotton cloth. Sometimes, the workers take coconut oil from their hairs in their index finger and clean the inside portion of rings whenever the breakages are high in a particular ring. Studies of M. Ramesh Kumar and M. Parthiban[9] indicate that 30–40% of hairiness could be reduced by lubricating the rings.

The selection of lift and ring diameter for the count of yarn spun is very important. The spindles revolve to introduce twist in the yarn. As the bobbins are fitted on the spindles, they also will be revolving all the time as the spindle is revolving. The yarn produced and wound on the bobbin also shall be revolving along with the bobbin as far as the spindle is revolving. If we have longer lift and ring diameter, the time taken to fill the bobbin will be more. The yarn that is produced and wound on the bobbin will have to make unnecessary revolutions along with the bobbin. The surface of the yarn will be getting rubbed because of the air resistance. Further to that the centrifugal force acting on the yarn surface shall be trying to throw the loose fibres out. This leads to higher hairiness. It can be seen clearly by observing a ring bobbin, that the bottom portion of a cop has more hairiness compared to top portion.

6.3   Barre

6.3.1   What is Barre?

Barre is a normal complaint received by a spinner, especially when he is supplying yarns to knitting T-shirts for export oriented units. The word "Barre" has been defined by different scholars and associations as follows:

(a) "An unintentional, repetitive visual pattern of continuous bars or strips usually parallel to the filling of woven fabric or to the courses of circular knit fabric".   – A.S.T.M

(b) "An unwanted stripe effect, regular to fading out, as a light or dark horizontal line".   – Z.T. Bartnik[10].

(c) "A consequence of subtle differences in yarn reflectance between individual threads in knit structure".

– J.W. Coryell and B. R. Phillips[11].

(d) "A continuous visual barred pattern or stripiness parallel to the yarn direction that is caused by physical, optical or dyeing differences in the fabric structure acting either singly or in combination to produce the barred pattern.   – Herbert T. Pratt[12]

(e) "A fault in Knitted fabric appearing as light or dark coursewise stripes and arising from difference in luster, dye affinity in the yarn, loop length, yarn linear density, twist, hairiness, etc.

– Textile Institute, Manchester

The barre is classified as simple, banded and complex. Simple is consisting of not more than two contrasting yarns, whereas a banded is a simple pattern with contrasting yarns alternate in equal width intervals. Complex is consisting of two or more interspersed simple patterns. The barre may be random, regular, distinctly visible, not distinctly visible, seen only after dyeing, seen in grey but not after dyeing, etc.

Although spinner gets complaints and claims from his customer relating to barre, in a number of leading show rooms and stalls at five star hotels, it is seen that T-shirts with distinct Barre are sold as "self designs" or "special effects" and they fetch higher price. The author has seen personally some customers insisting on that type of T-shirts.

6.3.2   Reasons for Barre

Barre is a result of variation, repeating at intervals. It may be from raw materials, spinning processes, winding processes, knitting, weaving,

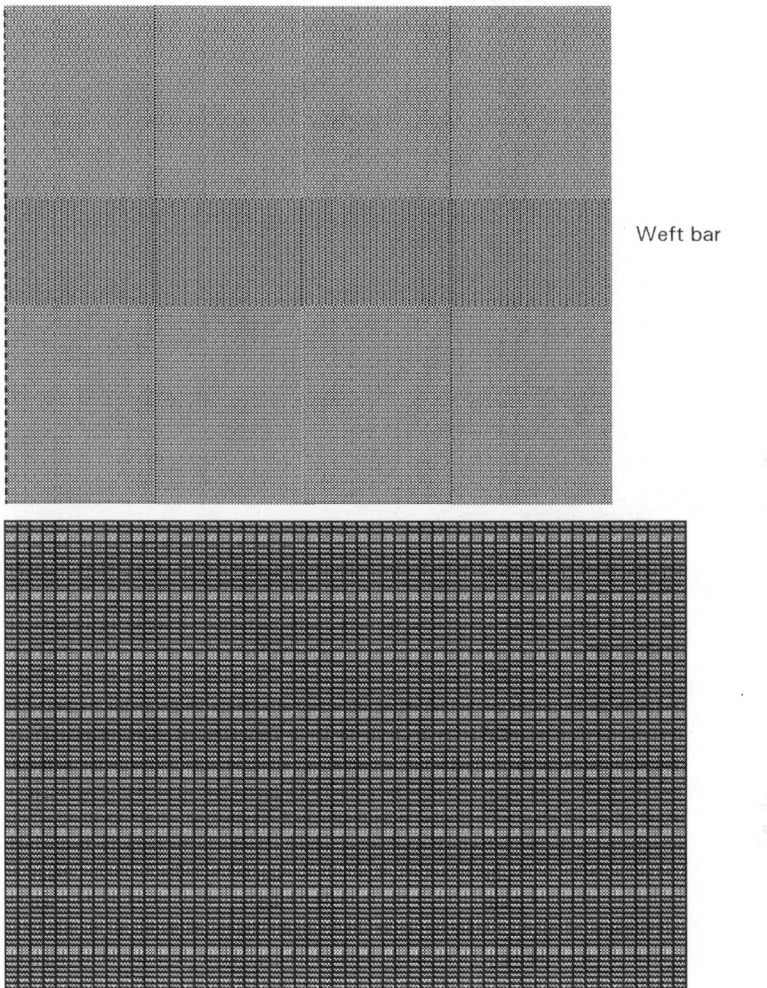

Weft bar

6.14 Barre effect on a cloth.

handling, storing, dyeing, etc. We shall discuss the raw material, spinning process and winding related reasons on which a spinner has control and he is responsible.

Raw material related

Although there is variation in raw materials, it cannot give barre unless otherwise the average fibre property of the mixing is shifting. The problem is more with direct bale feeding or with stack mixing. If the number of bales taken per cotton lot is more, then the lots shall run out very fast and we get variations in properties from mixing to mixing.

The variation in average value of micronaire value between mixing is more critical for barre compared to length or strength. It is always better to select cottons with consistency in micronaire value. Normally mills have a practice of checking the cotton properties as they arrive and record the readings. The mixings are decided by those readings. In a number of cases it is seen that the cottons are purchased in the season and stored at mills for a long time. The studies show that the micronaire of cottons shall be more in the fresh cottons and it gradually reduces as they are stored. This is because the cotton fibres had liquid in their lumen when they were in the plant and on the seeds, which gradually dried up. It takes three to four months for the lumen to completely dry. Hence the micronaire of stored cotton was found less by 0.2–0.3 in all varieties compared to fresh cottons purchased in the season. Therefore, the micronaire value at the time of preparing mixing is more important than the values recorded while purchasing.

The colour of cotton measured in terms of Rd and +b value is also an important factor contributing for barre. The Rd value should be with in a range of ± 2 to get consistency in colour. Here again we need to understand that the Rd value is affected by the dust particles on the surface of cotton which is a temporary phenomena. The draw frame sliver that is being fed to speed frames should be consistent in Rd value and +b value. The cottons should therefore the tested for Rd and +b value after cleaning thoroughly in a trash analyzer and in a mini card as running cottons up to draw frame and checking might not be practicable. However, checking and recording the fibre properties of the draw frame sliver on daily basis and plotting a graph shall help in understanding the trends. This can be used for deciding the optimum stocks that should be in pipe line.

The difference in UV values in cotton is another factor which can contribute for the problem of barre. It is found that the UV values increase when the cottons are stored in open sunlight compared to the cottons stored in closed godowns. In a number of cases it is seen that the cotton bales are stored in open area, and sometimes covered with tarpaulins or HDPE cloths. Theses covers can protect the cotton from rains but not from the heat of summer. The colour gets slightly faded because of hot summer compared to the cotton bales stored in good ventilated godowns. However, whether this factor directly contributes for barre is not established.

The opening of tufts is very important. If some bales are opened thoroughly and others are not, we do not get an intimate blending of bales. This leads to variation in yarn properties. This problem is completely solved after the introduction of bale pluckers, which open the cotton thoroughly to tuft size as low as 20 mg, and 40–96 bales are fed in a mixing giving a very thorough blending of materials. In addition, the installation of additional blending machines as multi-mixer or uni-mixer shall help.

If the yarn lots are of big size and we allot fewer machines for that count allowing more number of days to complete the lot, we invite the problem of barre. It is always better to complete the lot as fast as possible and avoid the natural variations.

When the cotton crops are getting changed, we need to ensure that all old crop materials are run out and the new is started. The customer should be informed about the change in cotton crop and pack the materials with different lot numbers.

We can get the problem of barre in synthetic fibre spun yarn also because of mix up of yarns from different merge numbers.

Spinning preparatory related

Variation in blending or mixing, migration or mix up of materials from other mixings, variations in addition of soft wastes into the mixings, mix up of carded slivers with combed sliver or vice versa are the main contributors for barre from the spinning preparatory side. The poor mechanical conditions in preparatory machines can lead to long-term periodic unevenness in spinning, that can give streaky or barre effect in knitted or woven fabrics. The condition of cans and bobbins are very important as they can give unwanted stretch and lead to uncontrolled fine yarns at random, thus resulting in streaks or barre. The controlling of stretch, monitoring of top arm pressure, selection of appropriate spacers and condensers, maintaining the machines in good condition, avoiding sliver splitting and handling damages, continuous monitoring of voltage and autoleveller working are very essential to come out of this complaint.

The hank variations in draw frame sliver leads to count variation in the yarn which is very critical for T-shirts. The variations in speed frames are critical to socks as well as T-shirts, whereas any variations in ring frames are very critical for socks. The controlling of stop motions and the time of action of the stop motion in draw frame is very important.

The variations introduced while doffing the bobbins manually in a speed frame can introduce barre lines in 1–2 m of knitted goods that will be at random.

Spinning related

The speed frame bobbins, if not used in time can become soft. The soft bobbins get stretched easily resulting in fine yarns. Therefore, one needs to always follow first in first out system while consuming speed frame bobbins at ring frames. If the inter bobbins are too big and touching each other, they give unwanted stretch till the bobbins become slightly small, leading to intermittent fine yarns. The maintenance of bobbin holders in good condition to give jerk free feed is very important to avoid streakiness.

Mix up of bobbins of different mixing or hank is one of the major reasons for the problem of barre. The mix up might take place because of improper colour codification, poor lighting, and ignorance of the workers or any thing.

Improper setting of traverses guide allowing the traverse to go up to the extreme end of cots gives periodic variations as the traverse bar move to and fro. The pressure variations between top arms can lead to count variations between cops. The channeled cots and aprons give periodic variations as the traverse shall be guiding the rove periodically into the channeled and good portions. The eccentric rollers in drafting give short-term periodicity which is very critical for socks. The loose tapes, jammed jockey pulleys, ineffective spindle buttons, half pressed spindle brakes, etc., are the causes for variation in twists, which results in differential dye pick up. The barre effect in such cases is seen after dyeing. Whenever a tape breaks, it is suggested to remove all the four cops and clean them to remove the low twisted portion getting mixed with running material. The tapes do not break suddenly. The tapes become loose or jerky for sometime before breaking. We do not know as to how long this loose tape or jerky tape was working before breaking. This is the reason that all the four cops are to be rejected and not to be allowed to go along with regular material.

Any breakdown in ring frame or any machine cannot happen suddenly. There shall be some wear and tear of the machine parts, jamming of parts, some jerky motions, some slowing down, etc. It is therefore suggested keeping the material doffed separately of the machine under break down and release it after checking the quality thoroughly.

In a number of times it is observed that the spinning mills have different types of ring frames running on the same mixing and count. As the machines of different technology or age run at different speeds, we get difference in the elongation properties and the yarn diameter, although the count and TPM are same. Mixing up of yarns from one set of machines from another set can thus lead to problem of barre. The spinner should ensure that the types of machines are fixed for particular range of counts all the time.

Even though we take care in packing materials separately as per the type of machines, change in mixing, change in lot of cottons, etc., they are likely to get used together by customer. Please remember that no knitter or weaver can use the yarns exactly as per the packing of yarns and completely run out a lot and then use a new lot of yarn, as the creel sizes are different and also the order quantity. There shall be some compromise in the work, but if it results in a barre effect, the spinner is blamed.

Winding related

In winding the system of first in first out must be followed strictly and no doff should be kept unutilized for long time. Mix up of old doffs and new doffs, even though of the same count can lead to barre problem, because of tension variations.

As the ring frame configuration can give differences in yarn diameter, hairiness, uniformity, etc., it is necessary to wind the yarns separately and pack them separately depending on the make and type of machines, even though the count, mixing and TPM are same.

Sometimes the cones become bad due to winding problems like stitches, EYC failures, sunken nose, ribbon formation, etc., and these cones are rewound. It is observed, because of the yarn moving in the opposite direction compared to yarns from ring frame cops, the rewound cones have always higher hairiness, higher twist and larger bulk. It is normal that a particular winder is allotted the job of rewinding, and the rewound cones are sent for packing. As these cones have a normal chance of getting packed together in a carton also go together in knitting. This leads to bands in the knitted fabric as all the cones from this carton shall look different from the cones creeled from another carton on the same knitting creel. Therefore, the rewound cones should not be packed and supplied along with normal yarn. They need to be packed as a different lot.

Any rough surface in the yarn path increases hairiness. Also the variations in wax application can lead to difference in friction and hairiness. The difference in friction value can lead to uneven loop heights leading to a barre.

In autoconers, after every break, the drum runs in reverse direction for sometime. This reverse movement disturbs the yarn surface. If the breaks are more on a particular drum, the yarn from that cone can give problems of bars. It is therefore very important to see that the breakages are as low as possible. We need to learn producing lesser faults in yarns rather than to depend on winding machine to cut the faults.

It is normally found that the remnant cops are fed to the last drums as they are nearer to the bobbin collection area. Because of this, that particular drum has to stop more frequently and produces a poor quality cone. We need to concentrate on spinning and ensure fewer rejections of cops at autoconers.

The winders are supposed to verify the labels and colour codes before putting the bare cones on machine. Any wrong labeling can lead to a mix up and barre complaint.

6.4   Classimat faults

6.4.1   What are faults?

The evenness testing indicates the variations in a yarn with reference to the average diameter. But we cannot really understand the nature of the defect or cannot conclude whether it is a fault. There shall be natural variations in diameter of yarn depending on the number of fibres in the yarn cross sections, but all variations do not create a working problem in the next process nor give a poor appearance that is unacceptable to customer. If the diameter is excessively high or low that can contribute for poor working or poor appearance is called as a fault; or else it becomes an effect. For example, we do not call slubs as a fault in a slub yarn, whereas it is a fault in a normal yarn.

6.4.2   Classification of faults

The working of the machines in weaving, winding and knitting depend mainly on the quality of yarn that is fault free. The size of the fault in terms of length and diameters is very critical in deciding the performance. The evenness testers, although indicate that a yarn is not uniform and has excessive variations, they cannot indicate the criticality of the fault. For example, if the diameter is above 50% of normal diameter, we get an indication of thick place, but we do not know the real diameter or the length of the defect. Similar is the case with thin places. Therefore, by referring the evenness readings, we cannot decide on the optimum settings that can be kept in a winding machine to clear the faults while getting optimum production efficiency.

The working of looms and knitting machines are greatly affected by the faults that have very large or very less diameter and also the higher lengths. Therefore, we need the data indicating the length and diameter of the faults and also whether the faults are coming at random or coming regularly. The concept of classification of faults got the recognition because of this.

The concept of classification of faults into 16 classes was developed by M/s Zellweger Uster, and the instrument was called as "Classimat". As the concept got acceptance, more classes were added. Now numbers of companies are offering their models with different types of classifications. The Classimat instrument was designed by Zellweger Uster of Switzerland and Classifault is developed at Japan by Keiosski. In Classifault more classes are identified, but as the measuring principles are different we cannot really correlate the figures.

Our objective of this paper is to discuss the reasons for these faults. The faults in the yarns are mainly due to improper raw materials, improper setting, worn-out parts and bad work practices including bad house keeping.

The faults are classified depending on the length and the diameter in a Classimat instrument. Figure 6.15 indicates the classification.

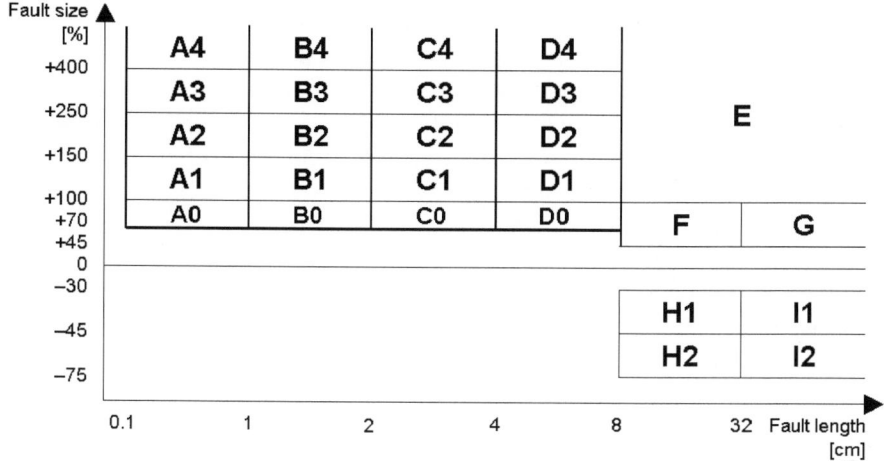

6.15 Classimat faults.

6.4.3   Sources of faults

The short faults are mainly due to improper raw materials. The improper raw material means variations in staple lengths, micronaire values, and improper blending. We get big slubs in case of occasional long fibres because undrafted ends. Normally they fall in C3 group. Bunch of short fibres, that are unopened give short length slubs of B2 and B3 type.

Variations in settings or improper pressures on the drafting rollers give periodic variations. Depending on the position where the defective part is situated, we get different lengths of defects. For example, an eccentric roller in draw frames give 10–13 cm repeats in draw frame sliver, but as it goes through speed frames and ring frames, gets drafted by 200–400 times. Therefore, we get long-term thick and thin places, but the same shall be insignificant to be classified as a fault in Classimat. If the eccentricity is in speed frames, then it gets drafted by 15–35 times at ring frames, and we get long-term faults of type G and I. If the eccentricity is in ring frame drafting zone, then we get the slubs of 3 cm length, but the diameter might be slightly varying; it means B1 or B2 faults.

The objectionable faults (D4, D3, D2, C4, C3 and B4) are mainly due to fluff getting contaminated, may be due to any reason. Sometimes, the fluff accumulated on the roof and creel fall on the running ends and gets twisted. We get Faults like C3, D3 and D4 because of such fluffs. The fluff collected in the drafting zone near the aprons, cradles, etc., contribute to faults like A4, A3 or B3. If the top clearer rollers are ineffective, there shall be occasional loose fibres going back and getting contaminated. This shall give A2 and A3 faults. Bad piecing or piecing from over the cots giveD2 and D3 types of faults. The broken end suction unit, if not clean, can result in lashing in. This can lead to C2 and D2 faults.

The studies of V. Ramachandran[8] show that long thin fault of H1, H2, I1 and I2 and the major six objectionable faults are also due to negative drafts between calendar roller and front roller as well as between creel and back rolls of a draw frame.

6.4.4   Misleading readings

Classimat readings may be misleading provided the person operating the equipment is not adequately trained. It is very essential to keep the EYC slots very clean. Some workers have a practice of taking the yarn through the EYC slot and wrap it on the cone before starting the machine. This is done to make the work easy. But this method gives wrong readings. The machine should be started first and then the yarn to be taken in each EYC one after the other. If the yarn is present in the EYC slot while starting the machine, the reference reading will be wrong, and we get abnormally high readings of I and H faults.

After each break, the faults are to be collected, and before starting the spindle, care should be taken to see that the EYC is cleaned with a cotton bud. If any fluff or yarn piece is remaining in EYC, it shall record as long thick place.

The trials made by the authour showed that the oil stains, if any, on the yarn is recorded as long thin places. This was confirmed by spraying lightly fine drops of coconut oil on the surface of cops and testing for Classimat after drying. Again it was reconfirmed by spraying fine mist of cotton seed oil. The cotton seeds that are not fully removed get crushed in the calender rollers of the scutcher in blow room. The seeds have oil and the same gets contaminated in the fibres. Theses fibres get spread. The yarn when tested show long thin places, whereas the yarn really does not have thin places. As the decisions are taken by the test results, we need to avoid this type of misleading results. We need to adapt sufficient gravity traps and remove the cotton seeds fully and do not allow it getting crushed. We need to educate the ginners also to avoid crushing of seeds but remove the seeds as a whole. The workers in mixing area and the blow room feeding area are to be educated to remove any oil contamination is visible. These are easily identified by installing U.V lamps on the long feed lattice and on both the sides of bale plucker (see Fig. 6.16).

6.5   Mix ups

6.5.1   Losses due to mix up

Mix up is a serious problem which is not accepted by any of the customers.

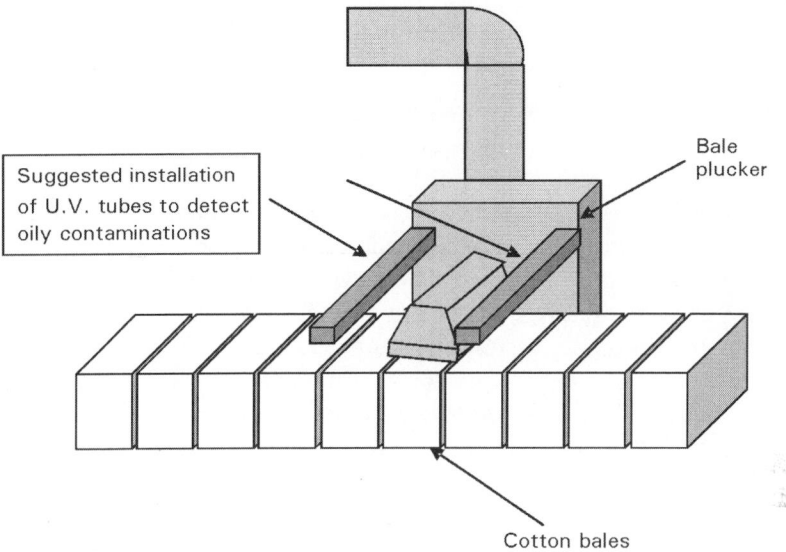

6.16 Installing UV lamps on bale plucker.

The mix up, if found, shall reflect very badly on the culture of the spinning mill. A customer can tolerate variations in quality but not the mix up. The losses due to mix up are very high. It results in high claims. The claim depends on the type of mix up. But customers do not prefer going back to same supplier if the mix ups are received in the supply.

The mix up is a result of poor management; may be in house keeping, training the people, planning the colour codification, planning the production activities, supervision, improper balancing of productions and stocks, excessive stocks, too frequent changes in counts and mixings, etc. Mix up cannot happen due to raw materials or technology. The technology and the raw materials can be changed fast by investing some extra amount, but culture cannot be changed over night. Therefore, the customers lose confidence on a supplier if they get mix up in the material supplied.

Mix up can take place at raw material stage, in between the processes or while packing. If the mix up is at packing, the customer observes it as a mix up, whereas raw material mix ups are seen as variation in quality. The mix up of yarns in cones is a serious issue, as it goes unnoticed up to the final fabric; it is seen only after dyeing or finishing the fabric. Even if one bobbin or a part bobbin gets mixed, it can create a big loss. The loss depends on the type of mix up. If the mix up is in warp, the loss shall be very high. One ring bobbin (cop) can spoil 3000–4000 m of cloth.

6.5.2   Precautions to be taken

It is very essential to follow channelization and colour codification to have proper control on the materials. Before allotting a colour code to a new mixing or count, care should be taken to refer the colour codes used earlier, and a senior man should audit all the areas and ensure that no material remains with the same colour code used earlier. The normal areas to be checked thoroughly are the quality control laboratory, sample department, the sider's bags of all the workmen, winding department, the cabins of senior managers, the waste cleaning area, the openings if any in return air trenches in winding and spinning sections, etc.

Before taking cotton bales for mixing, the supervisor or the mixing mukaddam should verify the bale numbers and lot numbers. All bales are to be kept at allocated place only. All laps, cans and speed frame bobbins are to be allotted the colour codes. The supervisor should personally check and ensure that the colour codification used is not confusing. Proper records are to be maintained for the colour codification used. Whenever a new count or mixing is being introduced, the colour codification record is to be checked. The codes should be kept constant for frequently running materials. This is very essential as number of workers have the habit of absenting for the work, and when they come for the work, just start doing the work without verifying changes in colour codification. On each machine the colour code is to be displayed by sticking a bobbin or colour strip.

Excess process stock is one of the main reasons for mix ups. If the stocks are kept low and the area is clean, the chances of mix up shall be less. People should be educated for not using a material if the coding is not proper or they are not clear of the codes used. The colour used should be distinct for the nearby counts. For example, if red bobbins are used for Ne 20^s, then we should not use any nearer colour for counts Ne 18^s, Ne 22^s, Ne 24^s, etc. If we use the same red colour for Ne 100^s, the problem will be less as workers can easily make out the difference.

In case of a count run out, one need to check all the machines – not only on the machines, but bellow the machine and also in the return air trenches, stocks not only in production area, but also in quality control area, sample rooms, cabins of spinning master and other senior managers, the waste cleaning area, etc. Once it is ensured that no material is remaining then green signal can be given to use the same colour code for a different count or mixing. It is always better that same colour code is not allotted at least for 10–15 days after a count is completely run out.

One of the major sources of mix up is while starting a new count on a ring frame. Once the bobbins are creeled and ends are taken forward, they need to be pieced with some yarn. As that count is new, people are forced to use some other yarn for starting. This shall lead to a mix up of 8–12 m of yarn in each cop. As this material is inside the cop, goes unnoticed to winding. If the counts are of near by counts, even the EYC in winding machine cannot catch the mix up. It is therefore suggested to cut the yarn in the bottom of all cops of the first doff with a sharp blade.

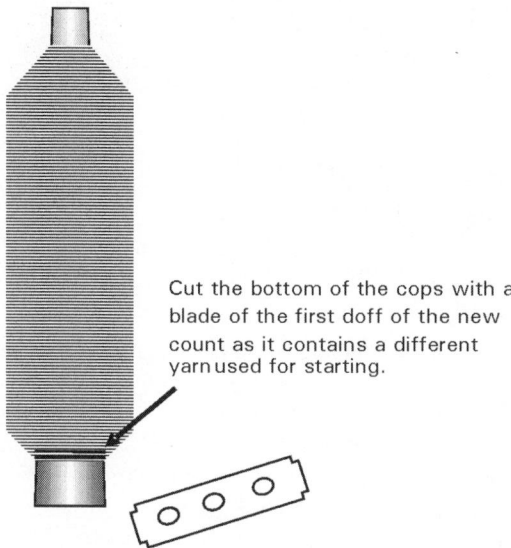

Cut the bottom of the cops with a blade of the first doff of the new count as it contains a different yarn used for starting.

6.17 Cutting of extra yarn.

Some mills keep specified colour bobbins for new count changing, so that winders also shall be clear that the yarns should break at the bottom of each cop and should not run out completely.

Checking of the cones under ultra violet rays are very essential to ensure that there are no mix up in the cones.

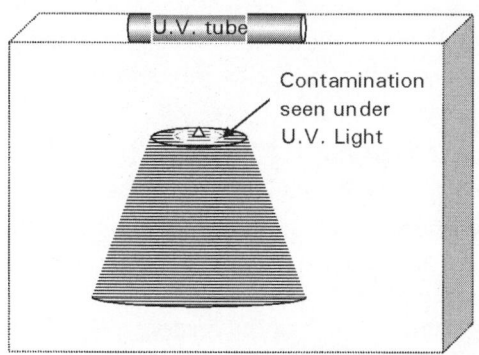

U.V. tube

Contamination seen under U.V. Light

6.18 Using UV light for contamination detection.

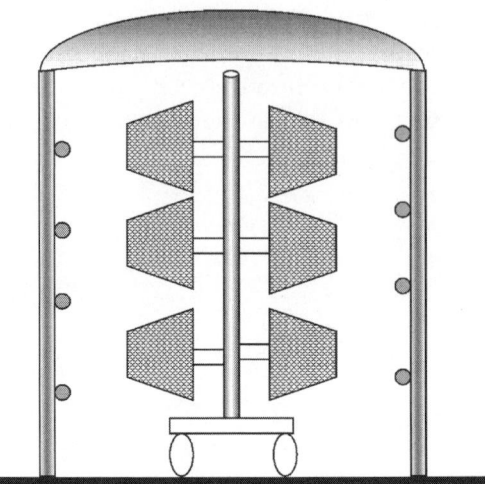

6.19 Cone trolley taken through a passage having UV lamps on both the sides.

The packing area should be secured and only trained permanent workers are to be engaged. There should be a record of the carton numbers packed by each packer or team of packers. It is the responsibility of the packers to check the labels, colour and codifications of the cones before packing. Recording of carton numbers against the packers shall ensure that the packers take full responsibility and pack only the correct materials. However, the packers cannot be held responsible if the labels are put wrong on cones in the winding itself.

In winding, before starting a new count all the old paper cones are to be removed from the winding machine and correct paper cones are to be put after verifying the labels. The supervisor should check this personally and give green signal for starting the machine. Do not encourage the workers to prepare empty cones in advance with labels fixed when you are planning for a count change. Avoid ad-hoc count changes. Make a plan in advance and inform all in advance about the count change.

6.6   Uneven cone length and higher breaks at tips

In earlier days, the customers were not insisting on uniform cone weight or uniform length, but were particular on the net weight of the packages. They were insisting on tail ends on each cone, so that they can keep the cones in reserve at a reserve creel. In case of a cone running out, automatically another cone was getting started because of the reserve creel. Therefore, stoppages of the machines were avoided. However, as the studies indicated that the tensions vary depending on the size of the cone and this variation in tension was a hindrance for high-speed working of warping, looms and knitting machines, the knitters and weavers started block creeling system.

Maintaining uniform tension on all the ends in a warping creel or knitting creel solve problems like Barre, breakages while working, etc.

The customer expects uniform length of yarn on all cones supplied to him, so that all cones can run out at a time without any leftovers on cones. If the length of yarn is not uniform on all cones, the moment the first cone runs out all other remnant cones are to be de-creeled. This results in higher wastes of costly and good yarn.

Earlier attempts were to produce cones of uniform weight. Some mills were employing people to weigh individual cones. The winder used to remove the cone from the winding machine and weigh for ascertaining the further winding to be done on the cone. He used to put back the cone on the same spindle and wind further to get the correct weight. Another method was to prepare a gauge by taking a cone of correct weight, and doff the cone by referring to the gauge (see Fig. 6.20).

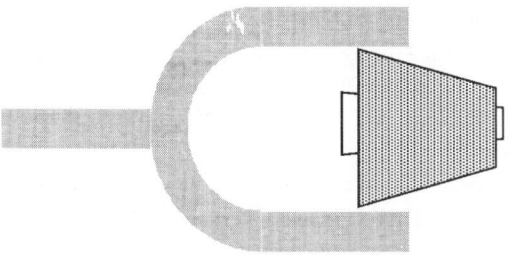

6.20 Cone diameter guage

Another method adapted was to fix a ring on the cone holder. As soon as the cone starts touching the ring, the winder used to doff it (see Fig. 6.21).

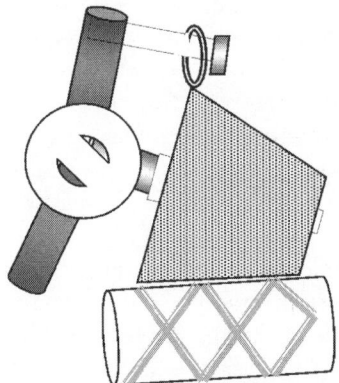

6.21 Ring fitted on cone holder to measure the cone diameter.

The use of gauge or the rings on the cone holders were helping to some extent in achieving uniform weights, but there used to be some variation due to differences in tension between drums. Although they were able to achieve fairly uniform weights, still the problem of warpers and knitters were not solved as the length were not uniform due to variations in count and also in the cone density. Therefore, the concepts of length measured cones were started.

Length measuring is done by counting the number of turns the drum made while winding yarn on a cone. People were happy with this development, but there were technical problems of the weight of the cartons. The carton weights were not the same as printed, and the governments were insisting on round figures in the net weight. With great difficulty, the mills were able to convince the trade about the length measured cones.

The studies done by the authour indicated wide variation in actual length of yarn wound on the cones compared to the length set. The variations were from 3 to 15% depending on the make of the winding machines. When a cone with a reading of 100,000 m was doffed and rewound on the same drum, it showed a reading of 88,000 m. There was no consistency in the length readings when all cones of 100,000 m were rewound again. The variations were from 85,000 to 104,000 m. Then a question was asked as to how the sewing threads were being sold on the length basis. To validate whether the sewing threads sold in the market have the actual length as stamped on them, yarns were purchased in the market of reputed companies. The studies showed that in sewing threads also the length was not remaining same on all packages. There were up to 13% shortage in some packages and 7% excess in some other packages. All the packages had sewing threads of three ply yarn wound on parallel tubes and there were no breakages while winding.

The variations in length in cone winding can be attributed to the slippage of yarn on drums while winding. The length is measured by the revolution of the drum with an assumption that the surface speed of drum is equal to the surface speed of cone. But as the cone has different diameters at the tip and at the bottom, it cannot have a uniform surface speed as that of a cylindrical drum with uniform diameter (see Fig. 6.22). Further, the variations were found more in waxed yarns than unwaxed yarns. Another reason found was the effect of breakages. Every time a yarn is broken, the drum takes a reverse turn and some yarn is wasted. If the breakages are more the variation in length is also more.

Because of the reason that the cone has different diameters, there shall be rubbing of yarn in the front, especially when the cones are small. As the yarn is wound up, the ratio of the diameter reduces. If bare cone has diameter of "d" in the tip and "2d" at the back, and yarn is wound of the size of "d", the ratio of diameter can be expressed as follows (see Fig. 6.23).

$$\text{Bare cone} - \text{Ratio} = d/2d = 1/2$$
$$\text{Full cone} - \text{Ratio} = (d+d+d)/(d+2d+d) = 3d/4d = 3/4$$

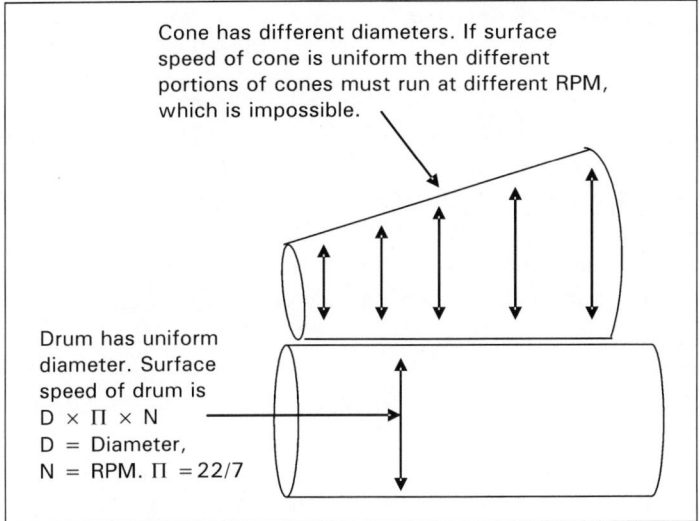

6.22 Speed of Winding

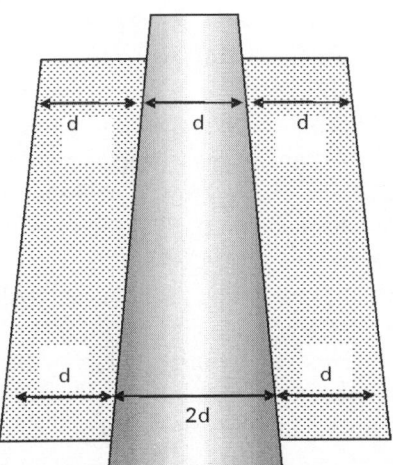

Fig. 6.23 Measurement of cone.

This phenomenon can be observed on the warping creel, that the creel breaks are more when the cones have become smaller than ¼ sizes. The inspection of cones, even under a magnifying glass cannot high light this problem.

The problem is more with synthetic yarns and where the angle of the cone is high like 9°15' compared to 5°57' cones. It is normally suggested to go for lower angle like 3°30' cones or 4°20' cones for high-speed warping.

6.7   Contaminations

Contaminations of dyed material in bleached varieties and white polypropylene or HDPE in dark dyed fabrics are normal complaints a spinner gets. There are also complaints of jute contamination, hair contamination and sometimes coir contamination. The decision of the rejection of the fabric or garment depends on the portion in which the contamination is seen, however, the spinner has no control on this. He needs to work for producing contamination free yarn. The steps normally taken for overcoming this problem are employing additional labour for manually sorting the contaminations, installation of scanners in blow room line that can detect the contamination and remove, educating the ginners to adapt better methods, insisting of covering the bales with pure cotton cloth and having contamination detectors at winding machines.

Manual sorting of contaminations by employing ladies was a normal practice during 1990s, which although costly was giving a good result, not only in getting a contamination free yarn but lower unevenness due to gentle opening of cotton by hand. However, this required huge labour and space; that was not affordable. After the introduction of Bale pluckers and contamination detectors like Vision Shield, Seccuromat, Vetal Contamination detector, etc., the manual sorting was reduced.

The spinners started insisting ginners to use pure cotton cloth for wrapping bales in order to avoid complaints of jute contaminations and HDPE contaminations. However, as the polyester started becoming cheaper compared to cotton, the ginners started using grey PC blends in place of pure cotton cloths, which became a bigger threat. This was first identified by V. G. Raghuveera et al.[13], who alerted all the spinners.

Laminated HDPE was developed to reduce the problem of HDPE fibring out and getting contaminated.

The problem of contamination is more if a mill is having both cotton and synthetic blends running side by side, producing fibre dyed yarns, improper house keeping and inadequate air changes in the production area. The workers wearing loose clothes or not covering their hairs and body properly also contribute for contaminations.

Process control studies

7.1   Process control

In the quality control there are two main activities. The testing of the materials being worked or produced is one part, which is called as testing and inspection. The second part is verifying the processes to ensure whether they are followed as per the requirement. The process control studies normally concentrate on the process. However, some of the process measurements are done with the help of inspection and testing also. Therefore, we are combining both in the discussions. The concept of process monitoring is that "If the process is followed strictly as per the requirement, we should get the results as anticipated".

Our objective is to design the yarns and produce to meet customer's needs and at the same time, the processes adapted should make manufacture and maintenance easy. The products are supposed to be made exactly and consistently to the specified design. Service delivery systems be planned to be consistent and reliable. In order to ensure that the processes are adhered, there must be performance measures and feedback on achievement. Proper analysis for the deviations should be made and actions are to be taken to correct the situation.

We have discussed various reasons for getting a poor quality in Chapters 4, 5 and 6. Knowing the reasons for poor quality is not enough unless we know the way of preventing it. Moreover, what exactly happened at that time is more important than all the possible reasons for a bad quality to generate.

The normal steps taken for manufacturing good quality yarns include planning the processes with suitable machines and technical parameters, selecting and procuring suitable raw materials and other accessories, having preventive maintenance ensuring good condition of machine parts, educating and training the workforce, etc. However, unless we are aware of what is happening, we cannot proceed further. We need to know whether the processes are as per the planned requirement. Process control is essential to prevent the poor quality.

In India ATIRA[14] (Ahmedabad Textile Industries Research Association) promoted the concepts of process control, which was supported and enhanced by SITRA (South India Textile Research Association) and BTRA (Bombay Textile Research Association). Various studies were designed in spinning mills to identify the deviations and taking suitable actions in time. The specific requirements of the customers, the feedbacks received, the performance studies, suggestions by the working people and benchmarking leading industries form the input for designing process control studies. Each mill shall have their own specific process control studies fitting to their requirements.

The process control studies normally include the performance assessment of the process or the machine being observed, deviations in set parameters, condition of various parts or sub systems of a process or the machine, the work practices likely to lead to poor quality and the quality or condition of the tools used for the process or maintenance of the process or machines. They can also be used for experimentation, assessing the performance of new trials or processes, etc. In the following section some examples of process control studies are discussed. In Chapter 8 several control points and check points are discussed specific to processes.

7.2   Mixing and blow room

7.2.1   Randomly checking the bales issued and bales planned

The mixing in-charge is supposed to check the bales and take in. However, in the absence of proper checking, there shall be a tendency of relaxing and leaving the responsibility on juniors. If there are some checks or audits, people shall be always alert. The QC investigator checks the mixing issue slip and the mixing data entry book. He checks the markings on the bales issued.

7.2.2   Checking the tuft size

Checking the opened mixing to ensure that big tufts are not put and soft wastes are opened and put properly is an important activity. Opening the mixing by hand can never give uniform tuft size as we get with Bale Pluckers. Further, the human being get tired after doing work for certain time, and they need rest. They tend to push unopened larger tufts in the mixing whenever they are in a pressure of completing the job in a certain time. Hence in process control studies tuft sizes need to be verified.

If we have Bale Pluckers, then also we need to check the openness and tuft size periodically to ensure that the machine is working as set.

SITRA has developed a simple method for checking the openness of the material at different stages of blow room. A fixed weight of cotton is put in a measuring jar of specified diameter. A standard plate is put on the top with a predetermined weight. The height of the material is checked. If the openness is more, the height shall be more (see Fig. 7.1).

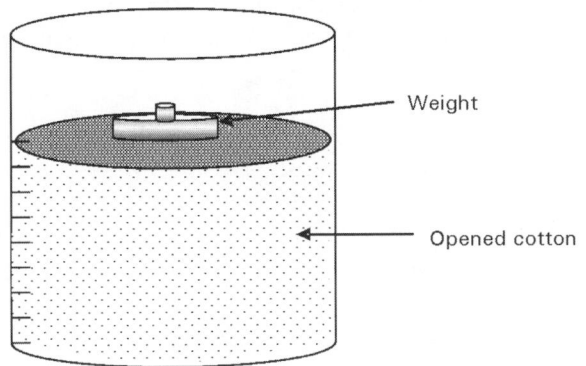

7.1 Fibre openness tester.

7.2.3 Synchronization study of different machines in a blow room

The production rates of back machines should be such that they work for at least 85% of the time when the scutcher or the card feeding units are working. If the machines are stopping for more time, it is an indication of wider settings. It is suggested to close the settings and get good opening. If they are running continuously, it is an indication of overloading of the machines. By checking the openness and the trash level, we can think of increasing the production in the back or else reduce the production in front.

7.2.4 Speeds of various beaters and fans

All beaters and fans are supposed to run at the planned speed, set as process parameters. There are chances of variations mainly due to slack belts, jamming of wastes in the beater sides and bearings and wrong pulleys. Whatever may be the reason, it should be corrected without delay.

7.2.5 Checking of air pressures, roller pressures at appropriate places

The quality of opening and cleaning depends on the various parameters set on the machines. Air pressures in the cotton conveying systems, pneumatic controls in the regulating devices, pressures acting on calender rollers, lap rollers, etc., are very important. They need to be verified periodically.

7.2.6   Grid bar settings

The grid bar settings are very important for getting proper cleaning and also in preventing costly cottons being wasted. The settings are decided depending on the cotton being worked. In a number of cases, especially when the mixing is changed, there are chances of this point being not attended. The QC investigator checks the grid bar settings and compares with the set standards.

7.2.7   Mass variation in lap sheet

The quality of lap is judged by the uniformity in the mass of lap sheet, length wise and width wise. Normally, laps are cut into 1 m pieces and weighed. After weighing 1 m bits, the bits are cut width wise and each bit is weighed. The comparison of weights is made as left, middle and right side of the lap, when seen from the front. The metre-to-metre lap CV% and also bit-to-bit CV% are worked out. By this we not only understand length-wise variations but also the width-wise problems. The length-wise variations are corrected by attending the piano feed motion and the synchronising the feed, whereas the width-wise variations are attended by adjusting the ventilation below the cages.

7.2.8   Total length of lap

Although the lap is doffed as per the length set, we need to verify at random whether the actual length of lap is same as what was set. In a number of cases we see some difference, mainly because of malfunctioning of proximity switches, difference in diameter of the measuring roller from what was assumed while designing the controls and stretch in lap after passing through the measuring unit.

7.2.9   Lap weight

Lap weight is normally the first control point of quality in number of mills from a number of years. Each lap will be weighed by the operator and shall reject the laps that are heavy or light beyond the specified tolerance. Correction shall be made if consistently the laps are heavy or light. Some mills had a practice of reserving some cards for running the heavy laps and some for the light laps in order to avoid lap rejections and reworking.

In recent years, because of the introduction of autolevellers in draw frames, the need for rejection of a lap is not felt. Also, by rejecting a lap we shall be doing more harm as the material has to pass through all the beating points again, that can generate neps and rupture fibres. Therefore, the system of rejecting the laps is gradually being discontinued. However, while setting a blow room for new mixing, the weighment is needed.

7.2.10 Cleaning efficiency

The main function of blow room is to open and clean the cotton. The effectiveness of blow room is measured by the cleaning efficiency. The trash content in a mixing is checked before being fed to blow room by using a trash analyser. About 1000 kg of mixing is taken and processed in blow room after ensuring the line was cleaned thoroughly. After running out 1000 kg, the wastes are collected from each machine and sent for weighing. Samples of opened material is collected from each opening point and checked for trash content. The cleaning efficiency is expressed as a percentage of trash removed to the trash that was present in the feed.

7.2.11 Machine audit

The points to be checked in periodic machine audits are the mechanical conditions of lattices, grid bars, beaters, swing door operations in hoppers, piano feed regulating mechanism, the pressure switches in case of chute feed system, the condition of bypass doors and rubber flaps, etc. Check the spikes on the inclined spike lattices, evener rollers and Kirschner beater, sharpness of blades in bladed beaters, etc. The pipe lines for transporting cotton pneumatically are to be checked and cleaned at regular intervals. The safety mechanisms like safety doors, switches, carbon dioxide flooding system needs periodic checking. In case of colour contamination detectors and metal detectors installed in the line, they need periodic calibration. The auditing points depend on the specific blow room line and the machines installed. We need to understand the activities of specific mechanisms and device system to study its effectiveness.

7.3.   Carding

7.3.1   Breakages and snap efficiency

The snap study means just counting the number of cards idle when a quick round is taken. We need to take rounds at different times and find out the percentage of machine stopped. Normally, snap efficiency shall be much near to the actual efficiency. For taking breakages, we need to study a particular card, either for one hour or for a full lap running time.

7.3.2   Neps removal efficiency and neps per gram in sliver

The neps removal efficiency depends on the neps fed to the card. If we feed more neps, then the neps removal efficiency shall be high. What is more important is the neps level in the out going material.

$$\text{Neps removal efficiency \% } = \frac{\text{Neps/g in material fed} - \text{Neps/g in delivery}}{\text{Neps/g in material fed}} \times 100$$

7.3.3   Card cleaning efficiency and trash in card sliver

The cleaning efficiency of a card depends on the trash fed. If we feed laps with higher trash, we get higher cleaning efficiency. It is important to see that the trash content in the out going sliver is at the lowest level rather than concentrating on cleaning efficiency.

$$\text{Cleaning efficiency \% } = \frac{\text{Trash\% in material fed} - \text{Trash\% in delivery}}{\text{Trash\% in material fed}} \times 100$$

7.3.4   Card-to-card waste variation

It is very important to maintain uniformity in waste extraction between cards to get uniform sliver. The QC investigator shall ensure that the cards are completely cleaned. Then the laps are weighed and fed to the cards. The wastes are collected from each point and weighed. The sliver produced is also weighed. Analysis of wastes like licker in droppings, flat wastes, cylinder droppings are made, and the wastes collected are analysed for trash content to ensure that good fibres are not lost in the droppings. Maximum difference of ½% wastes between cards are considered normal.

7.3.5   Trumpet size and hank of sliver

The trumpets should be decided on the hank of sliver and it should have a smooth surface. The workers prefer putting a slightly wider trumpet to prevent frequent choke ups.

7.3.6   Feeding consistency

In case of chute feed systems, we need to ensure uniform feed to all cards from the blow room. The level of cotton in the feed reserve boxes are to be maintained to get uniform feed all the time.

7.4   Draw frames

7.4.1   Periodic machine audit

Periodic machine audits, to ensure the machine is in good condition include, checking the condition of various parts like top and bottom rollers, the end bushes and the needle bearing, the gear wheels, the clearer strips, the trumpets, nylon mesh at suction box, etc.

7.4.2   Breakages

The breakages study gives a clear indication of the performance of the machines and the material. If the materials are uniform, the machine is in good condition and set properly, the breakages shall be less.

7.4.3   Top roller diameters

The top roller diameters are very important, as the pressure acting on the drafting has a direct link with this. We need to check periodically the diameters of all rollers.

7.4.4   Functioning of stop motions – delay in functioning, fail to function

The stop motion is very important function to be maintained in draw frame. It is not sufficient if a machine stops, but it should stop in time. It should not allow a broken sliver from creel to enter the drafting zone (see Fig. 7.2).

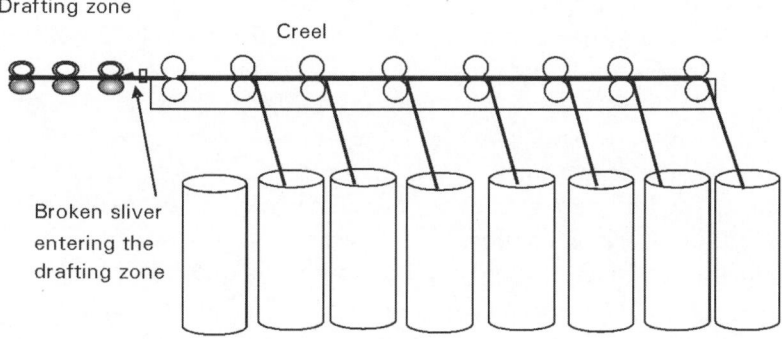

7.2 Broken sliver entering the drafting zone

We should understand that in case the broken portion enters the back roller zone, there is no correction possible, unless otherwise we have autoleveller. Even if the delay causes 10 cm of singles in the feed, it shall be multiplied by the draft, i.e., 8 at draw frame, 10–12 at speed frame and 20–35 at ring frame. Even if we consider lower drafts, then also the thin place in ring frame shall be equal to 10 ′ 8 ′10 ′ 20, i.e., 16,000 cm = 160 m of fine count by 14.3%. Therefore, in modern draw frames rapid stop motions are provided with pneumatic brakes.

7.4.5   Functioning of autolevellers

The functioning of autolevellers are checked in three conditions. First, by having all the slivers in the creel, then by cutting one sliver and adding one extra sliver, and then checking the weight per metre. If there are eight slivers in a feed, the testing shall be done with eight slivers, seven slivers and nine slivers. This is referred as "sliver test" or "A%" checking.

The autolevellers are sensitive to variations in voltage. Hence it is needed that they get stabilized power. It is necessary to monitor the voltage on a regular basis. "Over correction" and "under correction" are the terms used to indicate whether the correction is more than needed or less than needed. For example, when one sliver is removed in the feed, we expect the hank to become fine. Autoleveller when set properly will not allow the hank to change. If the hank is lighter, then we call it as "under correction". On the contrary if the hank becomes coarser after removing one sliver, we call it as "over correction".

7.4.6   Top roller pressure

Although pressure gauges are provided in the draw frames, the reading shown need not actually indicate the actual pressure acting on the top rollers individually. The pressure gauge shows the total pressure of air acting on the drafting rollers, but not on the individual pressure points. The pressure on top rollers can be checked by either by a Nip Load Pressure gauge or by carbon impression. The NILO metre, developed by ATIRA, Ahmedabad, is very popular among the mills in India for checking the Nip Load pressure. This gives the figure of actual pressure acting on the end bushes of a draw frame. However, this testing is done by stopping the machine, and putting a reference end bush instead of the actual end bush and the roller working on the machine. We can get a clear picture by taking carbon impressions. Normal method is to put one paper and take the impression of all the three rollers in one stroke. But in this case, we get information at one particular positioning. Hence, it is suggested to take impressions of individual rollers by putting a paper and carbon and inching the machine. By this we can get the information of two to three revolutions of a roller, which can clearly indicate variations in pressures might be due to eccentricity, uneven surface of cot, bowing effect due to excess pressure on both the sides, etc.

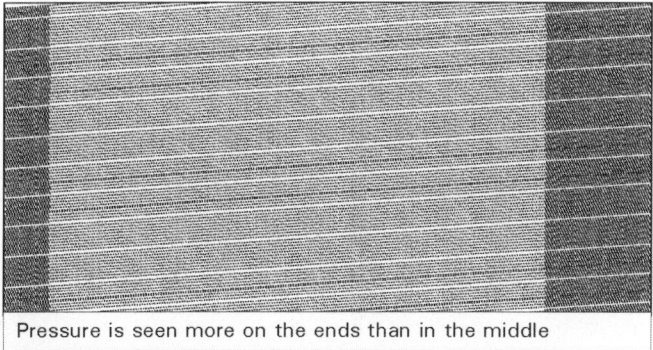

Pressure is seen more on the ends than in the middle

Pressure high at one side

7.3 Carbon impression method to check the pressure on single top roller.

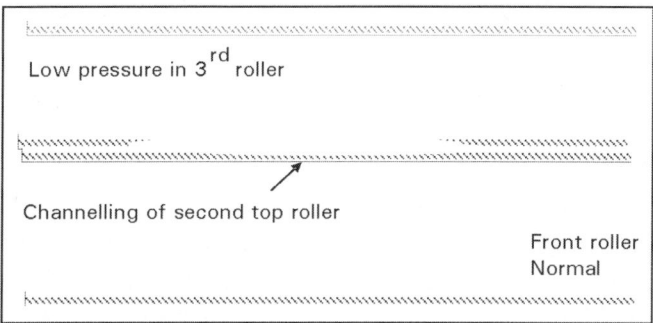

Low pressure in 3rd roller

Channelling of second top roller

Front roller
Normal

7.4 Carbon impression of all the rollers taken at one stroke.

While taking carbon impressions, we need to use fresh carbon paper all the time and discard the paper for this use. If the pressure is uniform, the impression will be uniform. If there is uneven surface in the rubber cot or uneven hardness in the rubber, we get different impressions and can easily rectify it.

7.5   Combers

7.5.1   Periodic machine audit

Periodically the condition of comber needles (half laps), the top comb, brush, nipper plates, detaching rollers are to be checked and decision is to be taken for procuring of the parts before they start producing bad quality.

7.5.2   Head-to-head noil variation

The noil variation between heads should be as low as possible to get consistent results. Periodic checking of variations in noil% and correcting any disturbance in settings is suggested. Care should be taken that the settings are not disturbed to make the noil% uniform. The settings on all heads should be identical. Normally, the productions and wastes generated for one minute is taken for this study. Higher head-to-head variations and machine-to-machine variation in noil extraction lead to unsatisfactory yarn appearance and higher breakages in spinning. The increase in average noil% increases the cost of manufacture.

7.5.3   Breakages

The breakages in comber should be aimed at zero. The QC investigator studies the comber breaks for full running out of one set of lap, and give the report. The reason-wise analysis is to be made and suitable action taken.

7.5.4   Short fibre removal and improvement in mean length

The purpose of combing is to remove short fibres of length below a specified limit and to improve the mean length of fibres in the sliver. The short fibres in the lap and the short fibres in sliver are checked using a suitable instrument. The short fibre removal percentage is worked out by comparing the short fibres in lap to the short fibres in the sliver and expressing the difference as a percent of short fibres in the lap. The improvement in mean length is the difference of mean length of fibres in sliver to the mean length of fibres in lap expressed as a percentage of mean length of fibres in lap. Generally lower short fibre removal and poor improvement in mean length results in irregular and weak yarns and also to higher breaks. However, we should know that when the feeding material has low short fibre content, the short fibre removal and improvement in mean length will be lower. Therefore, the level of short fibres and the mean length of outgoing sliver are more important than the efficiency worked out.

7.5.5   Neps removal efficiency

The neps in laps and sliver are checked using a suitable instrument. The neps removal efficiency is determined by the difference in neps level between the lap and sliver, expressed as a percent of neps in lap. Poor efficiency of neps removal results in poor appearance of yarn. However, we should know that when the level of neps in the fed material is low, we get poor neps removal efficiency. Therefore, the level of neps in out going sliver is more important than the neps removal efficiency.

7.6   Speed frames

7.6.1   Periodic machine audit

Periodic machine audit is done with a purpose of ensuring that all the parts of the machine are in good condition. A well-trained mechanic only can do this audit, as he can identify the worn-out parts precisely. As it is not practicable to check all parts thoroughly in a day, plans are made to check and verify one mechanism at a time when the machine is stopped for routine cleaning. The condition of each part is checked and the worn out parts are replaced. This study helps in proactively eliminating bad quality that could have come on a later date because of wearing out of the parts. This also helps in budgeting for the purchase of parts.

7.6.2   Breakages

The breakages indicate whether the quality produced in consistent or not as the rove breaks at a weak spot. It is suggested to take the breakage study for full machine and for the full doff, as the breakages vary depending on the bobbin position. It is suggested to mark the spindle number where the breakage took place, and analyse the reason for the break. If the back material is found as a reason, check the can to ensure whether the full can was bad or it was a stray incident. In case full can is found bad, remove that can and continue the study. Earlier, i.e., about two decades back, mills were happy if the breakages in speed frames were less than five per hour. Later studies revealed that each breakage in speed frame was contributing for multiple defects in yarn. This is because of jerks in all spindles due to inching to make the flyers aligned. Therefore, the present trend is to demand for zero breaks through out the working till doff. However, in case of some breakage, that spindle is kept idle till the running is complete and doffed and then the spindle is restarted.

7.6.3   Idle spindles

If any spindle is producing bad quality or having repeated breaks, that spindle shall be kept idle by the worker. An idle spindle is a loss as it shall be consuming power, labour, etc., but not producing anything. It is therefore necessary that a good mechanic discusses with the tenter, understands the problem and rectifies the mistake. In a number of mills it is seen that the supervisors and managers bring pressure on the tenter to run the spindle in spite of him explaining the problem. The worker then stops concentrating on the quality and just produces the number of hanks as demanded.

7.6.4   Stretch and draft

Periodically we need to check the hank of sliver fed and the rove delivered at different spindles in both front and back rows. By this we can find the actual draft and variation in draft between spindles. We also need to check the wrapping trend within a bobbin by taking the wrapping of full bobbin and plotting a graph. This gives an indication whether the stretch is increasing or decreasing as the bobbin is building up.

7.6.5   Evenness

Evenness checking is an important activity in process control. It should be done spindle wise and all the spindles are to be covered. Taking spectrogram and analysing it helps in identifying the sources of faults.

7.7   Ring frames

7.7.1   Periodic machine audit

Machine audit indicates the condition of the machine and a decision can be taken for ordering the spares in advance before a part completely wears out and results in a breakdown.

7.7.2   Breakages

The end breakages[15] in spinning are an indication of poor quality. We need to work towards zero end breaks, as each end break gives a piecing defect in the yarn, liberates fly and settles on the adjacent yarns being spun and converts good opened cottons into a wastes. The reasons for an end break might be the defect in back process, defects in spinning machines, working practices adapted especially while doffing, donning and restarting of ring frames, creeling systems adapted, wrong decisions taken by the worker and the technician while on work, atmospheric conditions, etc. The studies need to be done for full doff of a ring frame at different timings to understand the behaviour of the machine. Repeated Breakage Study (RBS) is a technique of identifying the rogue spindles that are giving repeated breaks and correcting them rather than just attacking on total breakages. The ring frame siders, while on work shall mark the spindle while attending the break and give the data to maintenance person in day shift for attending the spindle.

7.7.3   Idle spindles

If all spindles are working well and producing good quality, there shall be no idle spindle. Any worn out or missing part and disturbance in setting are the reasons for a spindle to remain idle. The sider keeps a spindle idle when he notices poor working or poor quality. Periodic studies are to be made reasonwise and the spindles are to be attended by an expert mechanic.

7.7.4   Ends down percent

Ends down percent indicates the percent of spindles not producing yarn at a given time because of a break. This gives a clear idea about the working efficiency and also the wastes being generated. The inspector should take a quick round and just count the number of spindles not working, without going for the details of the reasons.

7.7.5   Pneumafil waste percentage

If ends down give an indication of wastes that might be getting generated, actual weighing of wastes and expressing it as a percentage of production indicate the actual performance.

7.7.6   Spindle wise monitoring

Some mills have records of each spindle indicating the U% and imperfections. In case a spindle is continuously showing higher U%, then it shall be checked for the reason. In one of the studies done by the authour for identifying the reasons for count variation, the wrapping was done of both the feeding bobbin and the ring cop at the same time and the drafts were calculated. Even though the counts are varying between spindles, the draft should remain same. But it was found that drafts were also changing indicating problems in drafting, especially in the top arm pressure, condition of cots and aprons and the creeling stretch. This exercise helped significantly in correcting each spindle position. However, as it is a laborious process, continuous follow up is difficult. This system could be used when we get a higher count CV% within a ring frame.

7.7.7   Online monitoring of spindles

As a side effect of the development of information technology, now we have online monitoring systems of speed of individual spindles, breakages, the stoppages of the machine, etc. In a number of mills, where such latest gadgets are installed, people are happy in showing all that they have a system, but the actual analysis of data and taking immediate corrective actions as expected is found lacking. It needs thorough training of the staff and clarity on the priorities. As online monitoring systems generate huge data, one need to be very clear in identifying the priorities and taking action to correct the situation.

7.8   Winding

7.8.1   Breaks per cop

The breaks per cop are an indication of yarn quality including that of cop build up. In case of manual winding, we need a quality controller to conduct study by feeding 100 cops and noting down the reasons for breakages. In the automatic winding machines, which are very common now, the machine itself indicates the breaks per cop in each drum. We need to analyse as to why a particular drum should have a higher breaks per cop. We need to check the yarn path, the sensitivity of the electronic yarn clearers, the tensioners, the angle in which the cop is sitting in the feeding area, etc., and find out the correct reason.

7.8.2   Clearing efficiency

Electronic yarn clearers are provided on winding machines in order to clear the objectionable faults and long faults if any in the yarn. Periodic calibration of EYC is necessary to ensure that they are performing as needed. In order to assess the clearing efficiency, we need to check the Classimat readings of both cops and the cones produced from those cops. However, it might not be possible to check the yarns from same cops before and after winding, normally a full doff from ring frame is taken, and 50% is checked for Classimat faults before winding and 50% is checked after winding.

There were some efforts to rewind the same yarn after testing the cops for Classimat by taking due care that the faults are not cut in the first winding. However, it does not give the correct picture, as the yarns get reversed while rewinding and the size of slubs increase because of the rewinding.

7.8.3   Hard wastes generated

The hard wastes generated is an indicator of performance at winding and also the quality of yarn and the cop build up. Each break produces hard wastes of fixed quantity, i.e., the yarn required for knotting or splicing. In addition, if we have long thin and thick channels and off-count channels activated, the wastes generated shall be further more. In case a particular ring bobbin is repeatedly giving breaks, the same shall be rejected. The winder has to clean the cop and feed to winding again. If the contents left over on the cop are less than ¼, the winder normally do not use them. The senior officers should personally check the types of hard wastes and see that there is no unwanted generation of wastes.

7.8.4   Cone quality

The quality of winding, apart from the basic yarn quality, is very important to get smooth working in warping and knitting. The cones with uniform build, having no ribbons or patters, no stitches either in front or in the back and of uniform dimensions are the requirements of the customers. The process control investigator should verify the labels, the colour codification, cone weight, the cone density, the dimensions, etc., along with the quality of winding, i.e., without stitches, ridges, sunken nose, etc.

7.8.5   Length consistency

The warpers and knitters demand uniform length of yarn on the cone to have lowest residual yarns on cone while running out. The length measuring units are provided on winding machines, and the length measured cones are available. The studies conducted showed that none of the winding machines were giving the cones of the length set, and the variations were ranging from 3 to 15% between different makes of winders, all of international reputation. The difference was found even in the length specified sewing threads of reputed companies. The study was conducted by setting the length on cone winder and rewinding the doffed cones on the same drums and finding out the length. In another study, small cones were doffed with set lengths and the length was measured by unwinding the yarn on a reel. The reasons found were due to slippage of yarn while winding, tension variations while winding, reversing of drums while attending a break and absence of a validating system. It was a surprise to even machinery manufacturers that their machines were giving such a high variation in length. Efforts are needed to identify the contributors of length variations and attack them to reduce the national wastes, i.e., wasting good valuable yarns on all cones just because the yarn ran out in a couple of cones in a warp creel or knitting creel.

7.9   Packing

7.9.1   Calibration of balance

The yarns are invoiced as per the weight. The weight declared should be same as the actual weight. The customers also are very particular on the weight of the material they received as they pay for the weight invoiced. Further the excise and custom officials also are very particular about the weight of the materials. Therefore, the balances used are to be checked periodically and calibrated as needed by the trade and the regulations. Although a calibration frequency of one year is observed as common practice, because of the statutory requirement, we need to ensure that the balance is accurate at all the time. The packing in-charge should check the balance daily with a set of standard weights before starting the activity of packing. It is suggested to maintain a record of the actual readings. If possible, take a print out. By this system, in case of any deviations found, we would be able to call back that day's packing and recheck the weights. The quality control assistants can make random checking of the balances and the tare weights.

7.9.2   Cone inspection before packing

The packers are supposed to check each cone under UV light to ensure that there are no mix ups, check the labels, colour codification, cone density, cone quality, etc., before putting the cones in polyethylene bags. In some mills checking the moisture content of the cones ready for packing is also in practice. The process control team should check the effectiveness of UV lamps. The normal useful life considered is one year. If the bulbs are older, there are chances that the mix up might not be identified. The process control team also should verify the moisture metre used for assessing the moisture content in the cones. The labels are to be cross checked with the approved labels. The process control team also needs to check the strength of the cartons and the number of plies, the colour and shade of the cartons as well as the polyethylene bags.

7.9.3   Markings on the packages

The markings on the cartons or bags are to be cross checked with the statutory, legal and regulatory requirements, the agreement with the customer and the company norms.

Control points and check points

8.1 Concepts

In order to have best results, it is needed to identify the key result areas (KRA) in each process and monitor them. This involves checking and controlling. We need to identify the areas, that are to be controlled and that are to be checked to ensure proper control. The control points if controlled should lead to the achievement of the result in the key result area, and finally the company objectives and goals. The check points are process related, whereas the control points are result related[16]. General control points and check points are given in table below as guidance for production process. Department-wise points are given considering the general experience. This is only for guidance, and one can refine it considering prevailing systems and targets. It is necessary for the concerned heads of department to analyse the processes very carefully and decide on the area where they need control. Once the area to be controlled is clear, one can identify the points to be checked to verify whether the process is in control or not. Use of 5-S concepts helps significantly in deciding the control points and check points in process monitoring. This helps in identifying what is required to be maintained and kept, and what is to be discarded that becomes the basis for the design of work place layout and flow of process.

The normal control points in a production process are process parameters, selection of raw materials, selection and training of employees, maintenance of machines, rejection rates, delivery schedule, inventories, etc. For each control point, we can identify some check points. Here are some examples of check points for each control points.

Control points	Check points
Process parameters	• Level of adherence to process parameters • Calibration status of equipments monitoring the process • Method adapted for checking the parameters

- Suitability of process parameters to get the results
- The actual quality produced

Selection of raw materials

- Quantity of raw material received
- Quality of raw materials received
- Handling and storage systems
- Realization (output to input ratio expressed in percentage)

Selection and training of employees

- Competency levels of men available and men employed
- Process performance
- Work practices
- House keeping practices
- Discipline in work
- Reduction in absenteeism
- Increase in production per employee
- Increase in value addition per employee

Maintenance of machines

- Adherence to maintenance schedules
- Suitability of maintenance schedules and plans for the production and quality expectation
- Condition of machine parts
- Maintenance practices
- Results of maintenance
- Production loss due to breakdowns
- Consumption and costs of spares and lubricants
- Power consumption per unit production

Rejection rates

- Whether acceptance criteria are clear to all on the shop floor or not?
- Whether process parameters adapted are as per standards specified or not?
- Percent of process wastes and product wastes
- Percent of useable wastes and saleable wastes
- Rejections as a percent of input or output

	◆ Machine wise
	◆ Shift wise
	◆ Operator wise
	◆ Material wise
Delivery schedule	• Whether production started in time or not?
	• Utilisation percentage of machines
	• Productivity of each machine
	• Whether quality approved or not?
Inventories	• Material in process
	• Stock value of materials in stock at raw material stage, finished goods and material in process
	• Non-moving materials in stock

After deciding the check points and control points, it is needed to allocate the responsibilities for the people on the shop floor to check the process and take suitable corrective and preventive actions to bring the process in control and to maintain it. These points are to be reviewed periodically and modify as the systems improve. For example, in a scutcher at blow room, we get laps. We need to control the linear density of laps. Therefore, we fix a uniform length for all laps and physically weigh each lap. When we change the system to Chute-feed, there is no question of physically weighing the laps, but we need to ensure the required pressure and flow of materials. Therefore, we need to monitor the pressure switches. Similarly, we need to think of all the changes we are doing, may be as a part of technology upgradation or as a preventive action for certain potential problems or a corrective action taken considering the market feedback or by the trend analysis of the company performance. Extent of control depends on the expectation of customers, and hence, it cannot be same for all mills. Therefore, depending on the market feedback, one needs to continuously upgrade the control points and check points for his processes.

Any action taken to improve a situation should result in a change in the working system, and that change can be sustained only when we redesign our control points and check points and monitor the process as per the new requirement. Similarly any change in technological levels and in the work culture also initiates for a change in the list of control points and check points.

It is always desired to involve the men on shop floor to prepare control points and check points considering the process objectives. A brain storming among the concerned shall help in getting all the relevant points applicable for the process. It is further essential to pinpoint as to who should do which check, so that we can avoid duplication of work, and also, all shall be accountable for achieving the results. Proper work instructions are to be written for the system of checking, so that all can have the same system and the norms.

The check points should be written in short and clear wordings so that even a less literate person could also understand and implement without waiting for further instructions. It is essential to impart required training to the people assigned the job of checking, so that there shall be no bias, and people on spot get reliable inputs to take decisions relating to the corrective and preventive actions depending on the findings while conducting checks. The man who conducts the checks should be clear about the purpose of that check, so that he can be precise while making observations. If the objectives are not clear, there are dangers of just entering some data without proper verification. By this the decision or action taken shall be futile.

Frequent interactions with customers, both internal and external, shall help in defining the objectives of each checkpoint precisely. When we talk of check points related to each process in a production area, the requirement of next process becomes very important, as, if the next process gets the required quality and performance, the chances of failure shall be less with external customer.

It is essential to prepare control points and check points for each sub-process. It is necessary to fix responsibilities for each point so that they can be monitored efficiently. It is very essential to refer the operating instructions and the service manuals provided by the manufacturers, while deciding on the control points and check points relating to the machines involved. The control points and check points are to be referred regularly and use them as guidelines. These points shall ensure that, no vital point is left out in the process, which could have otherwise resulted as non-conformity or a complaint from customer.

The technical staff need to analyse the market complaints and other failures and find the root cause to help the operators involved. Once the root cause of a problem is established, the check points and control points can be decided to ensure the elimination of the problem noticed. Therefore, the control points and check points listed here are not the final one, but can be amended by the people on the shop floor considering the problems being faced and the targets fixed from time to time. One need to work out the priorities while deciding on the check points and control points, depending on the nature of problems or the feedback one gets from the customers. Check point, which was very essential yesterday, need not be essential today because of some modification in the systems or some change in the customer requirement.

It is essential to properly document these points and educate the users on continuous basis and audit the work to ensure that the same are being monitored.

8.2   Process-wise control points and check points

8.2.1   Mixing

Control points

(1) Selection of bales considering the parameters such as length, strength, fineness, colour, trash, maturity, neps, etc., and their variation to meet the quality of the yarn proposed.

    (a) Average micronaire of the mixing should be same for the entire lot. The difference in average micronaire of different mixings of the same lot should not be more than 0.1.

    (b) The micronaire CV% of a mixing should be less than 10%.

    (c) The micronaire range should be same. Cottons with too wide micronaire range should not be mixed.

    (d) Cottons with too wide reflectance value (Rd) and yellowness value (+b) should not be mixed.

(2) Deciding the proportion of different components in the mixing considering their properties, cost, age and stocks.

    (a) Immature fibre content should be as low as possible as it will affect dyeing and will result in white-specks.

    (b) Cottons with two different origins should not be mixed as far as possible, unless it is required to get a specific end result.

(3) Deciding the quantity of mixing to be done at a time.

    (a) If automatic bale openers are used, bale lay downs should be done properly, so that bales with different micronaire and colours are getting mixed up homogeneously even if small quantity is being checked.

    (b) If manual mixing is carried out, bales should be arranged and mixed properly so that bales with different micronaire and colours are getting mixed up homogeneously even if small quantity is being checked.

    (c) For manual mixing, the tuft size should be as low as 10 g.

(4) Issuing mixing in time.

(5) Thorough opening and homogeneous mixing.

(6) Deciding the addition of spin-finish, hygroscopic and/or antistatic agents, tinting colours, etc., depending on the materials being used.

(7) Adequate conditioning of mixing before feeding to the blow room.

(8) The work allocation for employees.

Check points

1. Whether the bales received are as per the plan or not?

2. Whether the floor is cleaned properly before laying the mixing or not?

3. Whether bales are opened gently or not?
4. Whether the Hessians and Bale hoops are removed and kept at designated place properly or not?
5. Whether surface of bales are cleaned before opening or not?
6. Whether cotton is opened and broken into smallest tufts as explained or not?
7. Whether the contaminations are removed by proper checking or not?
8. Whether the antistatic agent/spin finish/tint, etc., are properly prepared or not as per plan and applied uniformly?
9. Whether the layers of opened material are put properly or not in the stack to have a homogeneous mixing?
10. Whether the identification board is updated or not?
11. Whether the temperature and humidity are as per the requirement or not?
12. Whether the required time is allowed for conditioning or not?
13. Whether the men employed are as per plan or not?
14. Whether the required quantity of mixing is prepared or not?
15. Whether the usable wastes are received with proper identification or not?
16. Whether the wastes are properly checked and cleaned before adding in the mixing?
17. Whether the quantity of useable wastes added is maintained uniformly in all mixings or not?

8.2.2   Blow room

Control points

(1) Selection of opening and cleaning points and their sequence as per the mixing.
(2) Selection of process parameters like, speed, setting, hank of material delivered, length of lap in scutchers, etc.
    (a) If the micronaire is low, blow room process parameters become very critical.
    (b) It is better to do a perfect pre-opening and reduce the beater speeds in fine opening. If required one more fine opener can be used with as low as beater speed, instead of using very high speed in only one fine opener.
    (c) If the micronaire is lower than 3.8, it is not advisable to use harsh beating machines like porcupine opener, Creighton opener, bladed beaters, CVT4 or CVT3.
(3) Engaging trained workmen and providing additional training as needed.

(4) Evolving and implementation of maintenance schedules.
(5) Providing and maintaining safety gadgets as required.
(6) Deciding the work allocation per employee.
(7) Deciding the frequency and systems for waste evacuation and transportation.
(8) Deciding suitable identification systems for wastes and good material delivered.

Check points

(1) Whether all the cleaning and opening points required are in working order or not?
(2) Whether the bypasses are done as per plan or not?
(3) Whether the synchronisation of working of machines is suiting the quality requirement?
(4) Whether speeds of the beaters, fans, feed rollers, etc., are as per plan or not?
(5) Whether the settings are as decided or not?
(6) Whether the lines are cleaned or not as per plans and before changing the mixings?
(7) Check the droppings for presence of good cottons.
(8) Whether the hank of lap or weight per metre of sheet is as per plan or not and whether they are uniform?
(9) Whether the quality is produced as required?
    (a) Neps increase in cotton after blow room process should be less than 80%.
    (b) Fibre rupture in blow room should be less than 2.5%.
    (c) Lap weight, lap length, variation in yard to yard hank, uniformity of lap across the width.
    (d) Trash content and openness of the opened material.
(10) Whether the workmen follow safety regulations regularly or not?
(11) Whether neps generation and fibre rupture are with in control or not?
(12) Whether the men employed are adequately trained or not?
(13) Whether wastes are removed in time and labelled properly or not?
(14) Whether the temperature and humidity are as per requirement or not?

8.2.3   Carding

Control points

1. Selection of process parameters such as card clothing, speeds,

settings, drafts, hank, etc. It is observed that 70% of the quality is achieved if wires selected are suitable for the product being processed.

2. Maintaining the schedules for preventive maintenance like setting, grinding, mounting, etc.
3. Engaging trained workmen.
4. Maintaining the required temperature and humidity.
5. Designing and providing safety gadgets.
6. Following suitable colour codification and channelisation.
7. Deciding the work allocation per employee and specific works for each employee.
8. Deciding frequency and systems for waste evacuation and their disposal and implementing them.

Vijayakumar[6] suggests studying of individual cards up to yarn stage regularly. If the quality is deteriorated by 25% from the average quality, card should be attended.

Check points

1. Whether the machines are in good condition or not?
2. Whether card wire points are sharp and clean?
3. Whether the settings are done as specified or not?
4. Whether the wheels put are as per requirement or not?
5. The quality of web.
6. The breakages and their reasons.
7. Whether the wastes are removed in time or not?
8. Whether the temperature and humidity are as per requirement or not?
9. Verify whether laps fed are of good quality or not and there is no licking.
10. Verify whether the hank of sliver produced is as per plan or not.
11. Check the sliver weight difference between cards
12. Whether the sliver is uniform and U% is with in norms?
13. Whether the tenters carry out the work as specified or not?
14. Whether the machines and the materials are labelled properly or not for identification?
15. Whether all stop motions are working properly or not?
16. Whether the production obtained is meeting the targets or not?
17. Whether the card wire points are in good condition or not?
18. Whether the card is giving the cleaning efficiency as required or not? If kitties or seed coat fragments are more, higher flat speeds should be used and as much as flat waste should be removed to reduce seed coat fragments in the yarn.
19. Whether the neps removal efficiency is as per requirement or not?

20. Whether the trash and neps in sliver are within control or not?
21. Whether the fibre rupture is within tolerable limits or not? Uniformity ratio of sliver should be better than raw cotton as some short fibres are taken out in flat strips.
22. Whether the cans and springs used are of required quality?
23. Whether the wastes are removed from spring bottom before feeding cans to cards or not?
24. Check the cards for cylinder loading. The cylinder loading should be nil. If cylinder is loaded, wire should be inspected. If required, grinding should be done or wire should be changed.
25. Calendar roller pressure should be same in all the cards and sliver diameter difference should be less.

8.2.4   Drawing

Control points

1. Deciding process parameters, viz., settings, draft, number of doublings, speed, hank and length of sliver in can.
2. Deciding colour codification and channelisation.
3. Engaging trained workmen.
4. Evolving and implementing maintenance schedules.
5. Deriving and providing required humidity and temperature.
6. Deciding the work allocation per employee.

Check points

1. Whether the settings are as per requirement or not?
2. Whether the conditions of the machine parts like drafting rollers, cots, end bushes, saddles, hosepipes, springs, belts, bearings, etc., are good or not?
3. Whether the cans and springs are in good condition or not?
4. Whether the wheels are put as per calculations or not?
5. Whether the hank of sliver produced is as per requirement or not and the variations are with in limit?
6. Whether the machine is running at the planned speed and giving the required production?
7. Whether all the stop motions are functioning properly or not? Check for the response time after a break and verify whether it is stopping before the broken sliver enters the back zone in drafting.
8. Whether the voltage variations are with in control for autoleveller draw frames or not?
9. Whether in the sliver test A% is with in norms for autolevelled material or not?

10. Whether the colour codification and channelisation are followed as per plan or not?
11. Whether the workmen employed are adequately trained or not?
12. Whether the quality of cans and springs are as specified or not?
13. Whether the cans are fully cleaned before putting in the machine or not?
14. Whether the coiling is proper or not?
15. Whether the trumpets are of correct size and have smooth inner surface or not?
16. Whether the surface of the sliver table (creel) is smooth or not?
17. Whether the scanning rollers adapted are of correct size or not?
18. Whether the temperature and humidity are maintained as per plan or not?

8.2.5   Combing

Control points

1. Selection of process parameters, viz., half laps, speeds, settings, drafts, hank, etc.
2. Maintaining the schedules for preventive maintenance like setting, brush mounting, buffing, etc.
3. Engaging trained workmen.
4. Maintaining the required temperature and humidity.
5. Designing and providing safety gadgets.
6. Following suitable colour codification and channelisation.
7. Deciding the work allocation per employee and specific work for each employee.
8. Deciding frequency and systems for waste evacuation and their disposal and implementing them.

Check points

1. Whether the machines are in good condition or not?
2. Whether half lap points are sharp and clean or not?
3. Whether the settings are done as specified or not?
4. Whether the wheels put are as per requirement or not?
5. The quality of web.
6. The breakages and their reasons.
7. Whether the wastes (noil) are removed in time or not?
8. Whether noil% is as per norms or not?
9. Whether the top combs are cleaned periodically as per schedule or not?

10. Whether any long fibres are going in the noils?
11. Whether the table trumpets are as per norms or not?
12. Whether the temperature and humidity are as per requirement or not?
13. Verify whether laps fed are of good quality or not and there is no licking.
14. Check the hank and verify whether it is as per plan or not.
15. Check the sliver uniformity and U% and compare with norms.
16. Whether the tenters carry out the work as specified or not?
17. Whether the machines and the materials are labelled properly for identification or not?
18. Whether all stop motions are working properly or not?
19. Whether the production obtained is meeting the targets or not?
20. Whether the neps removal efficiency is as per requirement or not?
21. Whether the neps in sliver are within control or not?
22. Whether the cans and springs used are of required quality or not?
23. Whether the wastes are removed from spring bottom before feeding cans to cards or not?

8.2.6   Speed frame

Control points

1. Decide the process parameters, viz., settings, speeds, draft, hank of rove and twist.
2. Package parameters, viz., coils, taper, lift, diameter, etc.
3. Deciding on colour codification and channelisation.
4. Engaging trained workmen.
5. Evolving maintenance schedules and implementing.
6. Deciding and providing required temperature and humidity.
7. Deciding the work allocation per employee and specific works for each employee.

Check points

1. Whether the hank of rove is as per plan or not?
2. Whether the variations are with in norms or not?
3. Whether the stretch is in control in all the spindles or not?
4. Whether the breakages are within control or not?
5. Whether the U% of the rove is as per the requirement or not?
6. Whether bobbins are cleaned before putting on the machine or not?
7. Whether the bobbin hardness is uniform in all bobbins or not and within norms?
8. Whether the machines are running at specified speeds and giving the required production or not?
9. Whether all the parts of the machines are in good condition or not?

10. Whether the settings and alignments of rollers in drafting zone are proper or not?
11. Whether the drafting zone is always kept clean or not?
12. Whether the settings are as per plan or not?
13. Whether the top arm pressures are as required and uniform on all spindles or not?
14. Whether the spacers, condensers, sliver guides, false twisters, etc., are as per plan or not?
15. Whether the workers are following the work practices as specified or not?
16. Whether the workmen are adequately trained or not?
17. Whether the tenters and doffers are wearing sider's bags or not?
18. Whether the safety switches are acting or not?
19. Whether the house keeping is in order or not?
20. Whether the works are allotted as per plan or not?
21. Whether the colour codification and channelisation are followed as per plan or not?
22. Whether the maintenance is carried out as per plan or not?
23. Whether the separators are in their respective positions or not?
24. Whether the transportation of bobbins to ring frames is as per plans or not?
25. Whether the temperature and humidity are maintained as required or not?
26. Whether the house keeping is as required or not?
27. Whether sufficient empty bobbins are available or not?
28. Whether Bobbins are cleaned properly before putting on the machine or not?

8.2.7   Ring frame

Control points

1. Deciding and adapting process parameters, viz., count, twist, speeds, settings, draft, travellers, spacers, chase, winding and binding coils, cop diameter, etc.
2. Engaging sufficient and trained workmen and specifying the works for each employee.
3. Deciding and providing required humidity and temperature.
4. Deciding colour codification and implementing.
5. Deciding channelization.
6. Deciding the work allocation per employee.
7. Evolving the maintenance schedules and activities and implementing them.

Check points

1. Whether the ring frames are working as per the count pattern decided or not?
2. Whether the bobbins fed are as per plan or not?
3. Whether the count of yarn is as per requirement and the variation is within limit or not?
4. Whether the TPM of yarn is as per requirement and the variation is in control?
5. Whether the machine speeds are as per plan or not?
6. Whether you are getting the required production and efficiency as per plan or not?
7. Whether the machine settings are as per the parameters planned or not?
8. Whether the condition of the machine parts is good or not?
9. Whether the yarn uniformity, appearance, hairiness, etc., are as per requirement or not?
10. Whether the cop build is as required or not?
11. Whether the breakages are in control or not?
12. Whether the idle spindles are in control or not?
13. Whether the ends down are in control or not?
14. Whether the men allocation is as per norms or not?
15. Whether the workmen are adequately trained or not?
16. Whether the house keeping is in order or not?
17. Whether the maintenance is done as per plans or not?
18. Whether the travellers used are as per norms or not?
19. Whether the travellers are changed as per schedule or not?
20. Whether the siders are cleaning the draft zone as required or not?
21. Whether the bobbins used are cleaned properly before using on the machine or not?
22. Whether the bobbins are pressed properly or not before starting the machine after doff?
23. Whether the Pneumafil wastes are removed in time as per schedule or not?
24. Whether the colour codification and channelisation are followed properly as per plan or not?
25. Whether the wastes generated are with in limits or not?
26. Whether the siders are wearing sider's bags or not?
27. Whether the bonda wastes are promptly put in sider's bags or not?
28. Whether the wastes are disposed with proper accounting and labelling or not?
29. Whether the drafting zone and yarn path are always kept clean or not?

30. Whether the temperature and humidity are maintained as per need or not?

8.2.8   Cone winding

Control points

1. Selection of suitable process parameter considering the customer requirements, viz., speeds, yarn clearer settings, tension, cone dimensions, wax quality and settings at contamination channel.
2. Cone identifications: cone tip, cone label, winder number, machine number, lot number and contract number.
3. Engagement of trained workmen.
4. Evolving maintenance schedules and activities, and implementing them.
5. Deciding the work allocation and production targets.

Check points

1. Clearing efficiency of yarn clearers.
2. Production efficiency.
3. Winding drum-wise performance.
4. Winder-wise production and wastes generated.
5. Whether the paper/plastic cones used are of required quality or not?
6. Cone quality.
7. Weight variations between cones.
8. Hard wastes generated and the reasons.
9. Splicing quality.
10. Friction value for waxed yarns.
11. Cone identification.
12. Work practices.
13. Cleanliness and house keeping.
14. Condition of machine parts.
15. Cone hardness.
16. Increase in hairiness and imperfections after winding.
17. Working of all stop motions.
18. Whether the spinning bobbins have the correct identification as decided?
19. Whether the workers are using the spinning doffs with the machine traceability in view or not?
20. Spinning doff stocks and winder allocation.
21. Cones required in each lot to complete packing.

9

Role of technicians in quality management

9.1 Introduction

The technicians play a crucial role in quality management of any organization; spinning mill is not an exception. The technicians are the people involved in the design and establishment of the process, coordinating with the people on the shop floor and making them understand, implementing the process, monitoring and correcting them and achieving the required results in terms of quality and productivity. Therefore, the credit for success or failure of a process goes to the technicians on the floor.

Only by understanding the importance of technicians in process management, we shall not be able to do any thing unless we work out strategies for making a process success. The technicians are always under the pressure of producing the required quantity in time as per the agreed quality irrespective of the odd situations they face due to various factors relating to labour management, power shortage, maintenance lapses, non-availability of critical parts, changes in climatic conditions, sudden changes in customer requirements and so on. They have no right to give an excuse, but to give results under all circumstances by foreseeing the problems in advance and taking precautionary measures. They have been respected for this ability through out the world. However, unless one applies his mind and work out solutions for various problems he faced, it shall not be possible for him to be successful as a good technician.

Let us now study the steps in quality management in which the technicians are involved.

1. Understanding the requirements of the customer precisely, i.e., the count of yarn, the purpose for which it is used, the expected quality parameters, the quantity to be manufactured, the date on which the yarns are to be despatched, the packing specifications, etc.
2. Understanding the company capabilities such as process capability, human resource capability, funds availability, infra structure availability, etc.

3. Understanding the legal requirements of the process for the products and services planned.
4. Designing the product, i.e., yarn.
5. Designing the process for manufacture.
6. Deciding the measuring and monitoring processes.
7. Working out the quality plans and the production programme.
8. Planning for the raw materials and other spares and consumables.
9. Procuring the required materials in time.
10. Planning the maintenance activities.
11. Tuning the machines as per the process designed.
12. Educating and training the men on shop floor on the production and quality aspects of the product and the process.
13. Allocating the suitable competent workmen for the skilled jobs.
14. Monitoring the activities of people and synchronising them to get the best possible results.
15. Guiding people in their works and resolve interpersonal conflicts with in their section.
16. Monitoring the process periodically to ensure its suitability.
17. Documenting the procedures adapted, actions taken on various problems and the results.
18. Reporting the activities suitably to superiors as well as to the next person taking charge.
19. Analysing the reasons for deviations in process and product performance.
20. Working out suitable corrective and preventive actions and implementing them.

By giving a big list like the above, one might think that the technicians will get scared, but in reality they are doing all those works on a regular manner. They are doing much more than listed. There might be differences due to individual's capability, which deviate from man to man because of the education, knowledge, skills, experience, maturity, etc., and if we can discuss these aspects in details, it might help to bridge the gap.

9.2   Routine and special activities

It is normal in any spinning mill, that the works are grouped as "routine" and "special" activities. The routine work is the one, which is done regularly without any deviation. Normally, these works are allotted to certain people, who are good followers and religiously follow the steps. They normally include the recording of attendance, recording of production, data of machine wise production, quality, wastes, speed, efficiency losses, labelling the products, supervision and house keeping activities, scouring of machines, replacement of lubricants, replacement of ring travellers, etc. The Special activities require creativity and thinking, as the jobs are non-repetitive in nature, for example some modifications on the existing machines, modification in existing systems, launching of new products, special trainings given for staff and workmen, fixing of new standards, etc. The technicians normally like this type of special jobs so that they can show their capabilities. Unfortunately, they forget that unless the routine works are done systematically, the special works done does not give results in large-scale productions, and in maintaining the systems. Therefore, the special activities should always ensure that the routines are not disturbed. The routine activities are the backbone of any successful organization, and the rigidity in following the systems give the result, of course, when the process is designed logically. A well-designed and followed routine works ensure the stability of the organization, and assures the quality and productivity all the time. Let us take some examples.

1. Reducing the imperfection in yarn

The imperfections in the yarn can be reduced by various means, like introducing a better cotton with good maturity, uniformity in length and lesser trash, reducing the production speeds, increasing the saleable wastes at carding, blow room and combers, optimizing the settings, ensuring the proper maintenance of all parts in all the machines, maintaining the required temperature and humidity, educating the workmen periodically for good house keeping and work practices, selection of correct travellers, spacers, etc., replacing the card clothing, replacing of cots and aprons, keeping the drafting zone clean all the time, selection of suitable draft combinations and hanks, etc. Among the above activities some are "routine" and should be followed without fail. They are ensuring the proper cleanliness, maintenance of the machines, maintaining temperature and humidity, ensuring proper house keeping, supervising for good work practices, replacement of travellers, cots, aprons, in-time, etc. The act of selection of suitable raw material, settings, speeds, waste levels are all "special" activities, and need not be done daily. Even in spite of giving good cottons, good machinery, proper settings, etc., if we do not keep the drafting zone clean or do not remove the wastes from the machines periodically as decided, we can never reduce imperfections. We must first ensure that routine works are done religiously.

2. Improving the productivity

The productivity can be improved by modifying the machines to take up higher speeds, reducing the waste levels, reducing the stoppages due to various reasons, making the hanks coarser in spinning machinery, increasing the speeds, improving the raw material, maintenance of correct humidity and temperature, etc. Among the above, the selection of raw material cannot be done daily. Similarly increasing the speed is not a daily affair. But monitoring the stoppages, monitoring the wastes, monitoring the idle capacity are the routine works, which must be given priority all the times. If we do not manage these routine jobs, the modification of machines, giving a better raw material, etc., cannot give the required results. It might give a reverse result increasing the breakdowns, power costs, bad quality, etc., rather than increasing productivity.

3. New product development

New products are developed to have a competitive edge in the market. However, we should understand that once the customer accepts the product, we need to give that product in bulk, and, it no more remains as a new product. The following up of routine disciplines like maintenance, periodic checking of quality, following with workmen for house keeping and good work practices, maintaining the documents, following the codification system, monitoring of production, wastes, etc., are very essential for any product to remain in the market. There is no meaning in daily giving a new product to the market by making various permutations and combinations, unless we develop a system of maintaining the product developed.

4. Fixation of work norms

The fixation of work norms cannot be done as a routine job, but need to be done routinely on a periodic basis. No work norm can remain permanent in any organization. The job of a technician is to go on observing the working and the developments that are being taken place elsewhere and work out the possible improvements so that neither the man power nor the capital is underutilized. Sticking to the agreed work norms is a "Routine" work, and it should be followed religiously. If we deviate in routine follow up of implementing the agreed work norms, we will never be able to implement new work norms, however logical and economical it looks.

The above examples clearly explain the importance of maintaining the "Routine" activities. The success of quality management lies in maintaining the systems rigidly and not in developing new systems. But we cannot close our eyes for new developments. Any new development, once accepted should be made as a routine, and not to be treated as special all the time.

To have an easy monitoring of the routine items, it is necessary to identify all the routine activities and list them. Once the activities are listed, one has to identify the persons who should monitor, and also, the frequency of monitoring. They can be grouped as online monitoring, hourly monitoring, shift-wise monitoring, daily monitoring, weekly monitoring, fortnightly monitoring, monthly monitoring, etc., depending on the volume of operations, the works can be divided among different people and make them accountable.

Monitoring the activities are not just inspecting and testing, overlooking and ensuring the activities as correct, but also involves proper documentation. One should be specific in designing a document, as it should not be just filling up records by sitting at a corner and writing every thing as okay. Always record the readings and not the comments. These documents are meant for helping the industry to investigate and find root cause for the problems they face and not for satisfying an auditor or a manager. Avoid the comments as "Checked and found Okay", but insist on writing the actual readings. Also decide on what is required to be documented, so that it shall be a useful document for you for further developmental activities. Always ensure that whatever work you do is adding value and not increasing cost.

9.3   Understanding the requirement of a customer

Understanding the customer requirement precisely is the first job in any process. Unless we are clear in this we cannot give the required quality of product and services. The customer gives his requirement and expectations in a form, normally as a specification, but shall express his real concerns only when there is a problem. The purchase orders do not contain the concerns of a customer. It is therefore the technicians should take interest in going through the various feedbacks given by the customers. It is essential to segment the customers by the end use of the product, the region and the diversification they have. Although a particular customer has not expressed his concern for a particular product, we need to go by the concerns expressed by the customers in that segment and work for overcoming it. It is always better that a technician visits the customer to understand the technical requirements, rather than depending on the descriptions given by commercial persons.

It is a fact that customer normally gives his minimum requirements in writing, but shall be expecting a lot more from the suppliers. For example, no customer writes that packing should be attractive, but shall reject the material if packing is not tidy. It is the duty of technicians to find out the impact of each parameter at the customer's end. The technicians should discuss with their customers and prepare a list for their products, and educate the concerned shop floor people to monitor them. This exercise should be done customer segment wise. Let us see some examples.

- A weaver manufacturing fabric for defence end use shall be more particular about maintaining the average count at a certain level as the fabric weight have a tolerance, whereas a weaver making the same variety of fabric, but selling in civil market might insist on slightly finer count so that he can get more length of fabric, as the fabrics are sold on length basis.
- In some countries the weather is always cold, whereas in countries like India, it is normally hot. When yarn goes from a hot country to a cold country, its property changes, as the natural wax adhering to the cotton becomes brittle. Even within India, some places are always dry and some are always humid. The customer shall be specifying his requirements considering his weather conditions, whereas the supplier ensure those specifications in his working condition. Normally, the problem of weight shortage, excessive linting, loss of strength, etc., can be attributed to this. If we know clearly the conditions, we can always take that into consideration and design our products. The worst part is that the customers assume that the supplier knows his conditions, and the supplier always thinks that the working atmosphere shall be identical with that of supplier.

One more important factor is the language of customers. Although it is said that the communication is in a particular language, the exact meaning of that particular word used might be different depending on the culture and colloquial language in practice. Same word might give different meaning leading to confusion. Similarly is the case with workmen. Their language change from place to place and section to section in the same organization.

The requirements of a customer may be classified as quality, delivery, price, after-sales service and response time. Studies by experts have shown that 90–94% of customer grievances are due to areas other than product quality. It does not mean that quality of product is not important, but the customer assumes that the supplier shall give the quality as required. It should not be thought that technician is not responsible for complaints other than quality. The delivery depends on how we monitor the production and complete the lots in time as agreed with the customers. The price can be lowered if we are monitoring the wastes and wasteful activities and ensure that we produce the goods at the lowest possible cost. The after-sales service depends on the expertise of the technician who visits the customer and resolves the issues. The quick response also depends on how fast we respond to the enquiries by the marketing.

A good communication with customer relating to the orders, specifications, deliveries, complaints, etc., are very essential, and the technicians should take active part in understanding the real requirements of the customers.

It is also a fact, especially in spinning mills, that the market complaints shall be more when market is bad or when the rates are increasing. However, the technicians should work to eliminate those problems also, so that they can hold the customers with them for long.

There are various methods of understanding the customer requirements. Getting the competitor's samples and discussing on the improvements required is a better method rather than on discussing on the specifications and tolerances. We should know exactly as to what is the impact of different parameters.

In some mills, there is a practice of displaying the details of the market complaints received prominently in the production area and display the complaint sample also. They also display the name of the customers on each machine and the complaints received from that customer for the earlier supplies. The purpose is to educate the supervisors and the workmen relating to the precautions to be taken while in process. This system gives a clear indication to the shop floor people on to where they need to concentrate while producing an item for that customer. The role of the technicians is to follow up with the workers as well as with the seniors to ensure that the mistakes done earlier are not repeated.

Sometimes we need to manufacture the products without a firm order, and they shall be sold in open market. In such cases, we should target a segment and go by the feedback given by the majority of users. Also the spinner should have records of the best achieved situation and try to maintain it all the time.

The real task of a shop-floor technician is in translating the customer requirements into in-process specifications. This is discussed in length in designing a product.

9.4   Understanding the company capabilities

Understanding the company capabilities and communicating precisely to marketing can reduce the friction between the customers and the marketing personnel. Please remember that we are not politicians giving assurances to voters to get elected. We are a part of a supply chain, supplying yarns to our customers, who in turn shall convert them into fabrics and supply further. If my customer fails because of my mistake or false assurance it is a loss to our company. It is therefore very essential to understand our capabilities and limitations, and work on improving them.

The company capabilities can be studied in the following manner:

- Process capability, i.e., whether the process adapted at our mills can produce the product as per the requirement of the customer in terms of quality, cost and volume.

- Human resource capability, i.e., whether we have the required caliber of people to produce that product as required by the customer and to give after sales service as required.
- Finance, i.e., whether we can afford to hold stocks of raw material, packing materials, finished goods, spares, etc., so as to ensure uninterrupted supply to the customer as per the requirement.
- Environment, i.e., whether the present social environment is supporting to our business with the customer. This might include the trade regulations, pollution norms, political issues, power supply situation, draught or floods affecting the working, seasons affecting the working, etc.

The roles of technician is very important in working out our capabilities and inform marketing from time to time so that they do not give false assurances to customer, which can lead to problems later. There is no meaning in giving false assurances just to sell one consignment, by which we might loose the customers permanently.

In order to have quick information, it is essential to collect data of each machine and each person and go on entering the data to get trends. We should be able to know as to which machine is good for which product, which worker is good for which product. Rather than knowing which machine or which man is good for a particular product or customer, we should have a clear knowledge as to which machine or man is not suitable for this product. This is more important as the market complaints are mainly due to such machine or workers.

The technicians cannot keep quite after identifying certain process or machine or man as not suitable for a particular product or customer. He has to work for improving the situation, as the management wants to supply that product to that specific customer, as it is profitable. So the main job of a technician is to work for improving the situation and make the men, machine and process as capable of supplying that material to customer. So they shall have to continuously work and give proposals to top management and improve the systems. Remember, others cannot do this job.

9.5   Understanding the legal requirements of the process

It is very important to identify the legal implications of a process and take suitable proactive measures to adhere to the requirements. This includes packing norms, safety norms, pollution norms, employment norms, various factory acts, insurance acts, welfare acts, etc. It is the duty of the technicians to ensure that the legal requirements are met, as in case of any violations they are the one to be punished. You shall have to take the help of government recognized bodies for identifying the real requirements as per the law in force. In India, the Textile Commissioner's office and the Textiles Committee provide information to the mills through circulars. A number of Mill Owners Associations are doing a job of regularly collecting the information on the changes in regulations and send circulars to their member mills.

Make a list of applicable regulatory requirements for your section of operations, and indicate from where you can refer for the standards and norms. Have a record of fulfilling the requirements, which is normally called as a compliance record of your section so that you can monitor when some work is pending. Inform the concerned from time to time in case of any deviations found, in order to take corrective measurements in time.

The safety aspects cover not only the process but also the product. It is our duty to ensure that the products manufactured and supplied by us are safe for use at customers end. The quality of packing, the packing dimensions, the weight of packages, the handling systems, preservatives used, the mode of transportation, the other materials stored and transported along with these materials, the insurance covered, etc., are all to be considered and monitored by the technicians.

It is also essential for the technicians to follow up with the concerned HRD personnel relating to the periodic check up of their employees for medical fitness depending on the nature of work allotted to them. It is essential to identify the risks in the job, which can lead to sickness and hazards and take precautionary measures in time. Further a technician is supposed to find ways and means for reducing those risks by properly designing the process and monitoring it.

9.6   Designing the product

Designing the product is considered as the sole responsibility of technical personnel, although they depend on the inputs from other persons like marketing, purchase, human resource, costing, etc. The process of designing starts from the selection of designing stage, which can be done only by a technician. He knows by altering what, he can achieve which type of effect. His rich experience can always help in identifying the stage at which designing could be made more effective. Therefore, the technicians are expected to have calm thinking and able to recollect from their experiences.

One of the important works of a technician in designing is to provide correct inputs for designing, which include the process capabilities, process suitability, effect of the process on other products and services, the human resource competency requirement, various data of earlier trials and developments, present limitations, availability of required raw materials and spares, technical know how, skills, infrastructure, etc. Product designing is one of the toughest jobs, as it has to meet all the requirements of customer, regulations, etc., and should be able to be produced at the cost without hampering other activities. If the inputs provided by the technicians are insufficient, then the designer cannot design properly. Therefore, the technician needs to have an access to relevant data, and, capture them and provide to the designer as needed.

While designing a yarn, the technician has to imagine it as working at the customers end, and think of the probable failures, which can affect the performance. Work out the loss the customer has to face in case of each failure and work out the priority in terms of risks and cost. One can afford to have breakages on a loom or a knitting machine, as it can only increase the cost of operation, but a snap in a rope of a hoist, or a snap in the conveyors can become fatal, and even a single failure is not acceptable. Similarly there are different weightage for the problems a customer faces depending on the risk involved. Some mistakes can lead to closure of business itself whereas some are tolerable. A technician is therefore suggested to make a list as per priority of problems and design solutions and monitor those areas where the risks and loss are high. This technique is referred as FMEA, i.e., Failure Mode Effect Analysis. To be successful in this aspect a technician needs to exert and understand the working culture at customer's end, and the type of mistakes that can be done at customers end and prepare a suitable fool-proofing device in the design that is being made.

9.7   Designing the process

The role of technician in process design is very crucial, as the success of the product in the market, the economy of the company, etc., depend on the quality of the process. A good process aims at highest productivity with the best achievable quality at lowest cost. The process design involves understanding of the strength, weakness, opportunities and threats of each process, the interaction between processes, the linkages, balancing for production and quality, deciding of the machineries to be worked, the speeds, settings to be adapted, the recipe to be used, the productions to be targeted, the controls and checks at appropriate places, the reviews to be made, the targets to be achieved, the wastes to be generated, the generation and utilization of by-products, disposal of wastes, training to be given, monitoring the training activities, deciding of the documentation, planning for storage of material in process and the finished goods, identification system for materials in process and finished materials, system of handling, stacking, preservation and dispatch and so on. The product can be produced as required by various combinations of processes, but the one, which suits us best, should be decided considering the resources available.

The technician is the one who has the maximum knowledge of all the activities involved in process designing. He is the only man, who can forecast the impact of various parameters in the process design on the quality and productivity. Therefore, the technician needs to be always on his toes and improve his knowledge.

A process design requires a number of exercises of collecting the data of present system and analysing the present situation, working out various process combinations by designing the experiments, formulation and verification of theories for the cause and effect of process changes, balancing of the resources and infrastructure available to get the best result, identification of training needs and planning for providing training synchronizing with the implementation of process change, etc. The technician should keep his information bank always open, and ensure that it is always active. Knowledge cannot be put in a fixed deposit. It attracts higher interest in current deposits and loses value in fixed deposits.

The main activity in process designing is educating the people on the spot for successful implementation of the designed process. Here the role of a technician is very important. In a number of cases it is seen that written documents are provided for people on spot to follow a system, but this shall not be effective. Only by reading we cannot understand the systems. It requires some guidance, some practical demonstrations, on-the-job training and experience exchange. How much we learn by different means can be summarized as follows:

- Only by reading – up to 15.0%
- By taking guidance – up to 30.0%
- By seeing – up to 50.0%
- By doing – up to 85.0%
- By sharing knowledge and guiding others – up to 90.0%
- By practicing through out life – up to 99.0%

Therefore, the technicians should work practically and demonstrate to the people the working as per the new process designed.

9.8   Deciding the measuring and monitoring of process

The measuring and monitoring of the process is very essential to achieve the results as anticipated. But what is to be measured and how it should be measured is a pure technical job, and a real technician can only decide it. We should measure those areas where we can do something. There is no meaning in measuring something where we are not taking any action.

Measuring for academic interest is not the objective of a technician working in a mill. So we need to prepare the list of what are to be measured, and, what we intent to do with the data. Once this is clear, we can work out the accuracy required in the data collection. Depending on the accuracy requirement, the measuring tool can be decided or designed.

The technicians should first prepare a relationship diagram indicating the cause and effects, which helps in designing the measuring devices. He can always plan for integrating the systems to source data for different applications.

Once the data is made available for the concerned persons, the act of monitoring shall start. Here again the role of technician is important, as he knows the side effects of different monitoring actions. For example, to reduce the cost of manufacturing, one might think of reducing the comber noil in combers or increasing the speed of the machines. A technician knows as to what happens with this type of decisions. He knows the value of the work he is doing, and where necessary, he shall reduce the speeds, increase the wastes, but still brings profit by maintaining a good quality and consistency in production, which help in winning the customers. A good technician knows the efforts required and the result that could be got by monitoring, and shall suggest the area, which can give maximum benefit. There is no meaning in controlling the stationary, lubricants, etc., but one needs to control the realization of raw material, utilization of the plant capacity, etc. It means the knowledge of costing at each process element is essential for a good technician. Without proper knowledge of cost and effect, one cannot design the monitoring systems effectively.

The critical monitoring areas of a spinning mills are normally the raw material costs, the power consumption, the machine utilization, the operating cost per unit of production, the rejections after manufacture, process stocks, the production per unit, the machinery down time for various reasons, the material handling expenses, the packing and forwarding expenses, market feedbacks, customer satisfaction, adherence to the set procedures and systems, morality of employees, safety and environmental protection activities, training of workmen for good work practices, etc. In all the above jobs, the technicians take an active role.

The act of monitoring involves studying the trend, identifying the root causes for the trend, forecasting the change that can take place, taking a precautionary preventive action, re-devising the measuring system, involving the people by educating properly to maintain the process as needed, and finally ensuring that the results are obtained as per planned efforts.

The monitoring process may be on-line or off-line depending on the process. Nowadays almost all machines have one or the other online information providing systems. The basic problem is that the technicians have no time to see what information the machines give. If we have no time to see the reports, then there is no meaning in spending money for such system. The real worth of a technician lies in how best he makes use of the information available. Unless we device systems to prevent bad material from moving forward, there is no meaning in our systems. The technicians should give priorities for controls. There is no meaning of controlling every thing, as it adds to the cost, and does not give any benefit to the customer. For example, keeping very close settings in autoconers is useless unless we know how to produce a good yarn. Hundred percent inspection of finished material does not serve any purpose, unless the process is monitored effectively. Only technician can think and do this.

9.9   Working out the quality plans

The quality plans prepared consider the quality objectives of the product, the process capability and the quantity to be produced. Depending on the process capability, the machines, men and other infrastructure are allotted and the testing and inspection plans decided. As a technician knows about his machines, he can foresee the problems, and decide on the monitoring to be done and the suitable person who can handle the machine and process. A technician can work out various combinations for balancing the process, by altering the speed, the feed, the delivery, etc., so that he can get optimum utilization of plant and machinery. He can plan for having coarser hank in back process while working for a coarse yarn for a lower end use, whereas for a higher end use, he concentrates on the final product quality rather than just production of back process. He allots the machines considering the conditions and not by just numbers. For a non-technical man all ring frames are same, but for a technical man each spindle is different. Whenever he wants to conduct trials or wants to make comparisons, he maintains the same spindles. Similarly he knows the value and worthiness of each and every operation he does, and hence he is the most suitable man for making quality plans.

A well-designed quality plan makes the job easy for the shop floor supervisors, who have to just follow the process as per plan. The junior supervisors should be alert in identifying the variation in process while implementing quality plans. The role of junior supervisors is to provide the required data for planning, like the problems in each machine, the quality levels achieved machine wise, the problems faced during the shift working relating to working, absenteeism, humidity controls, shortage or excess of process observed during working, etc. They should observe the process carefully, and identify the potential problems so that suitable precautionary actions can be taken.

Analysis of the market complaints and their causes are to be explained to each and every workman involved, and their suggestions are to be taken to overcome the same. This should be considered in quality plans. The inspection and test plans prepared in the quality plans should have a direct link to the market requirements and feedback.

Once the plans are prepared, the concerned people in the organization like purchases, HRD, marketing, and quality control, etc., are to be communicated suitably so that they can work out their plans and act. Here again the technician plays a vital role.

9.10   Working out the production programme

The most important routine task of a technician is working out the production programme of the complete mills by balancing the production of each variety to meet the required delivery schedules. In quality planning, we discussed the activities relating to one particular contract, whereas here we are seeing the complete mill as one. We have limited machinery of different capabilities, and have to allocate them to have a balanced production. Here it is not only the balancing of production; it also includes balancing of men. As each contract is completed, a fresh contract started shall be with a different variety and hence the balancing of production is a daily affair, and hence is a routine job.

There shall be pressures from marketing to supply one particular variety early, because of which we may loose the productivity at other places, as the time required for different varieties are different. It is true for all the processes in a textile mill. In spinning area the production per spindle differs depending on the count and TPI for a specified speed. Hence it is not possible to have a fixed pattern unless that mill produces fixed varieties all the time without increasing or decreasing any of the products. It is not an impossible task as a number of mills are successfully managing to work without changes; however, the majority of the mills are flexible.

The balancing of production has to be done in all the shifts by the shift supervisors depending on the availability of men, machine, completing time of lots and contracts, stoppages due to unforeseen reasons like break downs, power failures, etc., and disturbance in working due to various reasons, quality rejections leading for reworking and so on.

The production planning needs to take into consideration the number of empty bobbins available in each colour, on which the number of machines that can be allotted for a count shall be decided. If it is not done religiously, we might end up with stoppage of machines due to empty shortages, allotting two colours for a given count or allotting the colours without ensuring that there are no remnants. This might leads to mix ups.

While planning the machines, care should be taken to avoid running of different materials and counts side by side to avoid problems of contaminations.

9.11 Planning for the raw materials, spares, consumables, etc.

The planning of the materials required for production is one of the major works of a technician. He need to plan in advance and order for the required raw materials, consumable items like ring travellers, wax, etc., maintenance accessories, lubricants, spares, packing materials, etc., which all depends on the customer requirements. However, planning is done much in advance before getting the contracts for manufacture. It means the technician should have data on the trends in customer requirements, which sometimes shall be seasonal. A timely planning and placing orders in time only can help the mills to run smoothly.

9.12 Procuring required material in time

Only planning has no meaning unless it is implemented. This holds good for procuring the required materials like raw materials, spare parts, packing materials, etc. The technician knows the manufacturing programme; accordingly he should follow up with the concerned purchasing people to get the material in time. The commercial people at purchasing cannot understand the importance of the items ordered; hence they treat all as same. For them the yardstick is the cost. So the follow up from technician is very important to get the required quality.

Once the material is received, the concerned technician should verify the materials and approve them for use. Unless the quality is acceptable, it should not be used.

The technician also has a role in vendor analysis. Different weightage for quality, price, service and delivery are to be decided by the user technician and not by the commercial person. The weightage depend on the role of that material in the operations to get the required quality, which is specific to each mill, depending on the technological level, location of the plant, the knowledge and skill of the people employed, the quality expectations of the customers, etc. Therefore, one should not refer to the weightage given by someone at some other mill, but work out for his mill.

The technicians can suggest alternate products that can be used in case of non-availability of certain ordered material. They cannot delegate this responsibility to commercial people.

Writing correct specifications of the materials to be ordered is the responsibility of technicians. The specification should be complete. In case of spares, we need to write the make and year of the machine, the part catalogue number, the drawing if any, and give samples if possible and feasible. In case of chemicals we should demand for safety data sheet from the suppliers. Wherever applicable, the National or International Standards are to be referred, e.g., IS/BS/EN/ASTM/ISO, etc. The technicians should also specify the method of inspection adapted for the received materials before they are approved for use.

The technicians should decide and prescribe the method of storing the received materials depending on their properties, frequency of using, safety aspects, etc.

9.13   Planning the maintenance activities

Maintenance activities are one of the crucial activities of quality management handled by technicians. The activities include understanding the machine features, identifying the critical parts, understanding the tuning operations of the machine, studying the life of various parts, forecasting the spares requirement on time, dismantling and reassembling of the machines for the purposes of scouring, cleaning, overhauling, etc., erection of new machines, shifting and re-erection of existing machines, ensuring breakdown free performance of the machines at optimum speed and productivity.

The frequency of maintenance activities are to be worked out by studying the condition of the machines and the working conditions. However, it is normally seen that the maintenance frequency is followed as suggested by a research organization or by the machinery manufacturer. It is not correct, as the maintenance requirements depend on the way in which we use our machines. Therefore, the frequency is to be decided by the technicians considering their working atmosphere in the mill and not referring to any recommended figures of either manufacturers or research organizations. They can be used as a broad guideline, but we need to work out our programme depending on our machine conditions, working conditions, quality demands, customer feedbacks, break down history, the speeds at which the machines are working, the products being produced and the rate of production, etc.

The maintenance job is a highly skilled job and is to be done with dedication. Any lapse in maintenance can result in bad quality and breakdowns that finally reduces the working life of the machine. The work requires lot of concentration and determination. The maintenance works should never be postponed to achieve higher productions. While planning maintenance, precautions are to be taken to have minimum stoppages of the machines. This can be achieved by planning the required parts, accessories, men, etc., in advance and combining related activities.

The maintenance workmen should have undergone required training, and should have the required knowledge of the machine. They should have the knowledge of basic tools and their application. A technician needs to train the workers and monitor the works. He himself should be able to demonstrate the work if he wants results.

It is always better to avoid, as far as possible, opening of machines, as every dismantling and re-erections are potential problem of bad reassembling. It is always better to monitor certain parameters and study the trends. Once the trend shows the signs of deterioration, then open the machines. The indicators selected should be such that they give an indication well in advance before the machine becomes bad. The examples could be increase in neps level, increase in imperfections, increase in fibre rupture level, etc.

Safety of the men and machines are to be given maximum importance during maintenance. The correct tools are to be used to avoid damages to the machines. Proper precautions are to be taken to protect the people working on the machines during maintenance. The electrical connections are to be completely removed unless otherwise it is required for the maintenance operations itself. Proper signboards and locking system are to be made to prevent the machines from being started by mistake while maintenance operations are going on.

A deep knowledge of the lubricants is a must for the maintenance person, as an improper lubrication can reduce the life of the machine, can increase power consumption and can create problems of staining. The system of storing the lubricants and protecting them from being contaminated with dust, water vapour, etc., is very important.

A good maintenance man knows the role of each and every part provided in the machine; and does not allow any part to miss or to be bypassed unless an in-depth study is made. He allows modifications only after ensuring that there shall be no adverse side effect because of modifications.

One of the very important tasks of maintenance is to maintain the maintenance equipments in good condition and get the measuring and monitoring systems periodically calibrated. The tools and equipments needing calibration are the gauges of different types used for setting the machines, the pressure gauges, tachometres, voltmetres, multimetres, measuring scales, weighing balances, eccentricity gauge, spirit levels, etc. The condition of the tools like spanners, wrenches, wises, hammers, screw drivers, cutting pliers, etc., should always be in good condition or else they might slip and create more damage. A technician should do periodic tools audit.

Wherever pressure vessels are used like air tanks of the air compressors, the auto claves for conditioning of yarns, etc., proper care is to be taken in bolting all the leaks, verifying the strength of the body and getting the servicing verified by competent authorities.

The maintenance work should help in keeping the area clean and tidy. Spilling of oils, grease, wastes, water, chemicals, etc., on floor is to be avoided where maintenance activities are being carried out.

Monitor the quality and productivity by taking studies before and after maintenance. Keep track of each machine by documenting the history of all changes done machine wise. This shall help in planning the activities and taking preventive measures.

Please remember that a new machine can give good quality and production for some days, but shall be the cause for major problems if not maintained well. But a well-maintained machine can never give problem and the company shall be safe.

9.14   Tuning the machines as per the process design

Maintenance can help in preventing the breakdowns, and bad quality due to worn out parts. But proper tuning of the machines, in other words the setting of proper process parameters such as speeds, settings, pressures, drafts, temperatures, timings of various motions, etc., are very important to get the correct quality and productivity. This is the prime responsibility of technicians. Even with latest machines we can produce bad quality if we do not know the parameters to be set. A deep study is required and one has to refer to previous settings, speeds and other parameters before making any change. If you feel change in parameters is essential, conduct study on a small scale, and change step by step. Do not change all machines at one stroke.

After making changes in parameters, take sufficient studies to ensure that the changes have given the required results, and you are confident of continued results. The documentation of process parameters should be done only after successful establishment of the parameters.

Before tuning the machines it is essential to ensure that the machine is in a good condition and do not have eccentric/vibrating/worn out parts, as they shall never give the correct results. It is the duty of the technician to ensure it.

9.15   Educating and training the men on shop floor

Educating the people on shop floor is very essential to ensure that the process works as per plan. This includes giving information of the process, product, quality requirements, the customer feedback, the production expected, the precautions to be taken, the checks to be made, role of each man in achieving the target and the general discipline of the company and the industry. It is very essential to monitor the work practices.

The technician plays a major role in educating the workmen on the aspects of process, and he is the only man considered to be competent in doing that work. Non-technical people (not by educational qualification, but by knowledge and experience) cannot explain and demonstrate the process to the workmen giving full information and clearing all doubts. Workers always respect and listen to a man who can work with them and guide them when there is a problem. Practical demonstration of a system is always welcome compared to classroom lectures or written instructions. Explaining in the language of workmen is required and not showing our command on a sophisticated language.

9.16   Allocating the suitable competent workmen for the skilled jobs

Just allocating a man for a job never helps. We need to allocate competent persons for the job so as to get maximum advantage. It is the responsibility of the technicians to prescribe the minimum competency level for each job. The competency is expressed in terms of education, skill, maturity and training. One who is having the required knowledge of the job, required skills for doing the job, required maturity for taking decisions, and has undergone the required training to do the job is called as a "qualified person". We should not confuse with the university degrees or diplomas when we talk of qualified persons for a job. The university degrees or diplomas give a confidence in the employer that "Such and such a person has passed such and such examination, and hence, he is likely to have the knowledge of such and such a process". However this might be a myth, as we have seen that a number of university-qualified people are not capable of doing the jobs compared to so-called uneducated people. Here the most important thing is the personal skills and capability and not a certificate. The technicians on the spot should observe the people working and judge their capability, and decide as to where they could fit best. A proper allocation of job reduces half the burden of the technician, and he can concentrate on other activities rather than following up with the people.

The process of identifying the minimum competency requirement involves breaking up of the job to small elements, and then, writing the requirements. Here is an example of a ring frame tenter.

Jobs done:

- Creeling the bobbins.
- Piecing the ends.
- Cleaning the drafting zone.
- Removing the Pneumafil wastes in time.
- Identifying the rouge spindles and informing the superiors.
- Identifying any deviations and inform the supervisors, etc.

If we consider the work of creeling the bobbins, the tenter should have knowledge of the count running on his machine and the colour codifications used in the speed frames. He should have the practice of using a bobbin holder, should be tall enough to reach the top of the ring frame creel for keeping the bobbins or to remove the bobbins. He should have the knowledge and practice of lifting the top arms, drawing the rove through drafting zone, etc. His education level should be sufficient to read the count board. His eyesight should be good to identify the different colours used for codifications and to identify the spindles where bobbins are exhausted.

The act of piecing involves identifying the broken ends, stopping the spindle by his finger, removing the lapping if any, taking the end through the lappet hook and the traveller, feeling the tension of yarn before piecing, exactly putting the tail end of broken end to the front roller nip and putting the bonda wastes in his bag. The tenter is supposed to replace the traveller if missing or burnt, cleaning the rings with his fingers or a clean cloth before inserting a new traveller, feel the condition of the ring and inform the superiors if found bad, remove the lapping on the clearer rollers from time to time, cleaning the spindle sides with a long round brush, collecting and keeping the pneumafil wastes at suitable location and so on. For this act the eyesight, finger dexterity, hand movements and a good practice of piecing the ends is very important. He should be able to identify the traveller number and its suitability for the count he is working. His practice of sensing the tension of yarn can tell the history of the spindle. He can recognize various defects or short comings in the machines like worn out rings, worn out traveller, worn out lappet hook, worn out separator, eccentric spindles, short of oil in bolster, loose tapes, improper pressure in top arm, etc. This skill cannot be taught in a college, but has to come by determined working and continuous application of mind while doing the job. A technician standing by the side of a ring frame and not having a practice of piecing can never be able to understand the problems in the machine as seen by a good tenter.

Cleaning the drafting zone is done when the ends are working. The skill of a tenter lies in cleaning the drafting area without cutting the ends and without introducing additional imperfections. Concentration in the work, skilful operation of taking a small stick and removing the embedded fluff from arbours, aprons, cradles, apron tension brackets (L brackets), necks of fluted rollers, Pneumafil mouth pieces, side of traveller clearers, etc., requires patience in doing the job. His knowledge on the role of each part in the quality and production of yarn plays an important role. Unless one knows the precision of the parts, he cannot develop that skill of handling them gently.

Removing Pneumafil wastes in time might look as an unskilled job, but an experienced tenter shall be fast enough to sense the urgency and remove the wastes in time. The time he keeps the door open has an impact on the suction pressure. However, the new machines which have rotary system for continuous waste extraction does not depend on the skill of the operative.

Identification of a rogue spindle is not a separate skill, but is a by-product of the piecing skill and experience of the tenter. As he feels tension, heat, vibrations and sound of each spindle during piecing and taking rounds, he recognizes rouge, which is different from others. His maturity in deciding a particular spindle as rouge is very valued. An inexperienced technician insists the tenter to run the spindle as the yarn works on that, but a matured one shall see the quality of yarn produced and its effect on the total lot. A good technician recognizes the maturity of a tenter fast. Similarly a matured tenter can identify the vibrations, worn out bearings, dry bearings, wrong meshing of wheels, deviations in temperature and humidity, change in cotton component, change in twist, speed, etc., and gives feedback to superiors in time.

By studying all the acts of a good sider, one can visualize the minimum competency required for allotting the job of tenter to a person. Similar is the case for any job. You just split the jobs into elements and work out what is expected out at each step, and to achieve that, what type of man is required.

Only by writing the minimum competency level, the job of technician is not over. He should identify the existing competency levels and group them in different categories, and plan for suitable action like training, job transfer, job modification, providing assistance, etc. The action taken should help us in achieving our targets.

9.17   Monitoring the process periodically to ensure its suitability

Monitoring the process periodically to ensure its suitability is normally called as process control and supervision. Here a technician verifies the process adherence to the planned system and takes suitable action in case of deviations. He shall have his own checklists or a checklist provided by his superiors, and verifies the activities. He shall not be able to check all, but shall have priorities to cover the vital activities. Depending on the changes done on that day or in the shift, he shall change his priorities. This includes production rates, wastes generated, quality of material in process, stoppages, house keeping, material flow, material handling, packing and storing, human productivity, team working, information gathering and maintaining records, maintaining the discipline while working relating to channelization, colour codification, material storing, labour allocation, grievance handling, etc.

The monitoring of process includes providing the required work instructions, ensuring suitable working environment such as temperature, humidity, air circulation, lighting, noise level, dust level, control of discharges, etc., following up with people to ensure that they are working as per plans, the production is coming as per the plans, the quality levels are maintained, the safety and regulatory requirements are fulfilled and taking suitable action in time for the deviations observed.

9.18   Documenting the procedures, actions and the results

The results got today should not be the end, but we should get the results continuously. A good technician shall document the happenings systematically so that he can identify the reasons not only for failures, but also for the successes. The documents should clearly reflect the works done, actions taken, changes observed, reasons investigated, the root cause analysis made, etc. Just filling the pages cannot be called as a record. A record is meant for taking suitable action and hence, there should be some system of easy access to the required information by way of proper indexing. The technicians should decide on what is to be recorded, who should record, when it should be recorded, and where it should be kept. It is necessary to train the people on method of collecting data and information and recording them, so as to avoid mistakes and loss of time and energy.

The technicians should fix the life for each record which depends on the speed in which technology is changing, the cycles in fashions and the yarn quality requirements, chances of going back to old systems, etc., and discard unnecessary records. Just keeping unwanted documents shall only add to the space and confusions.

So many times, the records are maintained as per instructions of certain senior person, but no review is made. The records are not seen by any body including the one who instructed to maintain, and the people maintaining the records also do not know as to for what purposes this record is maintained. This type of records increases the work and cost and does not add to the value. It is therefore suggested to periodically review the records that are maintained and rationalize them so as to have the required records only and discontinue the unwanted ones.

The records become redundant and useless because of the changes in technology, introduction of new product and new systems, discontinuing of certain processes, etc. Therefore, periodic review of what records are to be maintained is very important, and, it is the job of a technical person.

9.19   Reporting the activities

We may do any work, but we need to suitably communicate it to others so that they shall continue and maintain the same. Handing over of charge at shift end is one such activity by which we explain to the next shift man regarding the activities done and pending, the problems encountered and actions taken, instruction given by the superiors, and show the actual spots where the work was done and so on. Similarly we need to explain to our superiors regarding the productions achieved, production lost, quality levels achieved, the problems faced, actions taken, results of the actions, etc. One who can report well is always considered as an efficient person, as he has the facts clear which help the management in taking suitable action.

The level of understanding and expectations of superiors are different from that of colleagues, and hence, while communicating we should be careful and communicate in the same level. We need to understand the expectations and mental status, and suitably explain.

9.20   Analyzing the reasons

Whatever happened should have happened because of our efforts and not by chance. Hence, it is essential to analyze the reasons for deviations in process and product performance and take suitable actions. The analysis is needed not only for failures but also for the achievements. We need to identify the factors contributed for success and work to make them stable. The role of a technician is very important in this aspect. He should not have any bias while analyzing the facts. He should never come to a conclusion just by his experiences, but should verify the deviations from different angles and take suitable decisions.

The role of a technician in quality management is very important in maintaining the quality and productivity in a spinning mill. There are still a lot of activities which are not listed here, but the men on spot need to realize their role and take proactive steps. A technician is the heart of quality management, and the success of any process depends on the dedicated efforts of planning, coordinating, implementation, review and actions taken by the people on the spot.

Notes

1. B. Purushothama. Five Golden Questions – A self assessment Tool – Quality Update Nov 2007 - Indian Society for Quality.
2. B. Purushothama. Numerical Evaluation of implementation of ISO 9000 – published by Fibre2fashion – 7 Jan 2009.
3. ISO 9000 help – www.iso9001help.co.uk
4. Sead Jahic, IDEA, Consulting, Gracanica (BiH) – The Objective of product quality – www.idea.co.ba
5. Geocities – Rings and Travellers – www.geocities.com/.../ringtraveller.html
6. Vijayakumar – Combed yarn for Knitting –www.geocities.com/vijayakumar777/yarnquality.html
7. Dr. R. Chattopadhyay – Quality consideration in Blow room –NCUTE pilot programme 30th and 31st Jan 1999.
8. V. Ramachandran – Role of Drawing in controlling sliver quality – Journal of Textile Association, Vol 61 – Sep–Oct 2000.
9. M. Ramesh Kumar and M. Parthiban – Lubricated Rings reduce hairiness of yarns – Indian Textile Journal – Nov 2007.
10. Z.T. Bartnik – Faults in Knitted fabrics, their causes and cure – Textile Asia, June 1986.
11. J.W. Coryell and B. R. Phillips – Identification of Barre sources in circular knits – Textile Research Journal, Feb 1979.
12. Herbert T. Pratt – Some causes of Barre – Knitting Times. June 1977
13. V.G. Raghuveera, Basha, S.B. Deshpande and B. Purushothama – Studies on contamination of cotton yarn – 2nd Unit Level Conference, Textile Association (India), Ichalkaranji Miraj Unit – 16th Jan 1999.
14. A. R. Garde and T. A. Subramanian – Process Control in spinning – ATIRA Silver Jubilee Monogram – 1974.
15. B. Purushothama – The end breakages in Spinning; Its causes and remedies – Vastra – Volume 10, 1972-73
16. B. Purushothama – Guidelines for Process Management in Textiles – CVG Books, 2007.

Further Reading

J. M. Grover – Contaminations in Indian Cottons: Sources and Remedies – NCUTE Pilot programme 30th and 31st Jan 1999.

Dr. H. V. Sreenivasa Murthy and Dr. A. K. Basu – Optimization of Opening, Cleaning and Blending at Blow room – NCUTE Pilot programme 30th and 31st Jan 1999.

Dr. R. Chattopadhyay – Quality consideration in Blow room – NCUTE pilot programme 30th and 31st Jan 1999.

B. Purushothama – Linking exercises – A strong tool for Quality Auditing – Quality Update – Aug 2007 – Indian Society for Quality.

Vivek Plawat and A. R. Garde – Spinning Tablet I, Blow room – The Textile Association (India).

Dipali Plawat and A. R. Garde – Spinning Tablet II, Carding – The Textile Association (India).

J. M. Grover and A. R. Garde – Spinning Tablet III, Drawframes – The Textile Association (India).

Piyush H Shah and A. R. Garde – Spinning Tablet IV – The Textile Association (India).

M. C. Sood and A. R. Garde – Spinning Tablet VI – Ring Frames – Part I: Yarn Quality and Productivity – The Textile Association (India).